Essential Clinical Anatomy
of the Nervous System

Essential Clinical Anatomy of the Nervous System

Paul Rea

AMSTERDAM • BOSTON • HEIDELBERG • LONDON
NEW YORK • OXFORD • PARIS • SAN DIEGO
SAN FRANCISCO • SINGAPORE • SYDNEY • TOKYO

Academic Press is an imprint of Elsevier

Academic Press is an imprint of Elsevier
32 Jamestown Road, London NW1 7BY, UK
525 B Street, Suite 1800, San Diego, CA 92101-4495, USA
225 Wyman Street, Waltham, MA 02451, USA
The Boulevard, Langford Lane, Kidlington, Oxford OX5 1GB, UK

British Library Cataloguing-in-Publication Data
A catalogue record for this book is available from the British Library.

Library of Congress Cataloging-in-Publication Data
A catalog record for this book is available from the Library of Congress.

ISBN: 978-0-12-802030-2

For information on all Academic Press publications
visit our website at http://store.elsevier.com/

Working together
to grow libraries in
developing countries

www.elsevier.com • www.bookaid.org

CONTENTS

PREFACE

One area that has proved challenging to gain a greater understanding in medicine, surgery, dentistry, related health professions and the students studying within those specialties, has been the nervous system.

All too often, resources available in the field of anatomy, neurology and the neurosciences for the study of the nervous system are frought with complex detail and pathways which can appear daunting.

Therefore, the purpose of this textbook is to provide the key facts, in an easy to access format, for each of the major components of the nervous system without over complicating the anatomy. It takes the reader through, in a step-by-step process, the basics of the nervous system and the key divisions from both the structural and functional perspective.

Then, an overview of the essential antomy and function of the brain is provided. Following on from this, each of the brain regions are dealt with in turn in a little more detail namely the forebrain, midbrain and hindbrain.

A summary on each of the major divisions on the arterial supply and venous drainage of the brain is given, followed by the key relevant facts and figures related to the spinal cord.

The final two chapters are devoted to the ascending and descending tracts that occur with the brain and spinal cord, and also the key areas of clinical relevance that communicate with other regions of the brain and in and around the head and neck.

What differs with this text is that rather than copius amounts of detail, the essential anatomy is presented, frequently in tables. That way the key facst about each territory can be accessed immediately without having to refer to numerous parts of the text continuously.

Another key feature of this text lies in the fact that when each area is dealt with, the pertinent clinical applications are discussed and frequently hints and tips on how to examine these clinically in the patient. This means that all key information is readily accessible in the same region.

The combination of ensuring the relevant detail is provided without over-complicating the matter, combined with the clinical applications and details on how to clinically examine each region makes this text incredibly unqiue.

It is hoped this text will be the perfect revision tool for upcoming professional examinations, to have at the bedside, surgery or outpatient department, or simply to have at hand for quick access to the key anatomy, clinical applications and reminders of how to examine a patient's neurological system.

I really do hope you enjoy this book and find it a great companion during your studies and professional life.

Dr Paul M. Rea

ACKNOWLEDGMENTS

There are several people who I would like to thank in making this book possible.

First I would like to express my gratitude to Elsevier for having the time, patience and faith in me while putting this together. They have really been the backbone to helping me realize my dream in publishing this.

I would like to dedicate this book to my mother Nancy, father Paul and dearest brother Jaimie. Thank you for being there and supporting me throughout everything – I am so proud of you all! Thank you also to Jennifer Rea – my sister-in-law – but more like a sister to me.

I would also like to extend a very special note of thanks to David Kennedy for hearing every update as to the progress of this book, and coping with me! Thank you so much!

Thanks also to Christine Kennedy and Susan Kennedy who have listened to all the chat as this book was being brought together. I would like to extend a note of thanks also to Yvonne and Allan Hoffman.

Also, thank you to a dear friend who has gone but not been forgotten – Mark Peters.

Thank you also to Elaine Jamieson, Richard Locke and Leana Zaccarini for all the years of very special friendship.

Finally, thank you to a dear colleague and amazing mentor who has supported me from when I first started my career as an anatomist through to where I am today – Dr John Shaw-Dunn.

CHAPTER *1*

Introduction to the Nervous System

1.1 OVERVIEW OF THE NERVOUS SYSTEM

The nervous system is comprised of two parts: the central nervous system (CNS) consisting of the brain and spinal cord, and the peripheral nervous system (PNS), consisting of peripheral and cranial nerves as well as the motor and sensory nerve endings of those nerves.

1.1.1 Central Nervous System

The cells of the brain and spinal cord are called *neurons* (or neurones) and are specialized cells which can electrically conduct and communicate with nearby neurons. The purpose of these cells is to allow communication of the surrounding environments to adjacent cells to help regulate the internal environment of the body, and to respond to external stimulations. Supporting "glue" called neuroglia also holds the neuronal tissue together. Neuroglia provides nutrients to the neurons, maintains electrochemical stability, but also helps to defend the nervous system environment from attack by pathogens.

1.1.1.1 Brain

The brain is a mass of convoluted neural tissue and is referred to as the *cerebrum* (Latin: brain). The brain is a complex organ consuming approximately 15% of cardiac output, and can only survive a few minutes deprived of oxygen. If it is deprived of oxygen, death will ensue quickly. It is comprised of two cerebral hemispheres – left and right, and together these arise from the embryologic *telencephalon*. The cerebral hemispheres process information related to a wide variety of functions, and will be dealt with separately in later chapters. Between these two massive cerebral hemispheres lies the *diencephalon*. This is the *thalamus* and hypothalamus. The telencephalon and the diencephalon together form the forebrain – the first part of the brain.

Essential Clinical Anatomy of the Nervous System. http://dx.doi.org/10.1016/B978-0-12-802030-2.00001-7

The middle part of the brain, *midbrain*, or *mesencephalon*, structurally comprises the *tectum, tegmentum, cerebral aqueduct, cerebral peduncles* and many nuclei and pathways (or fasciculi). It typically deals with *alertness*, the *sleep/wake cycle, hearing, vision, motor function* and some *homeostatic regulations* like temperature control internally.

The last part of the brain is referred to as the *hindbrain* and is comprised structurally of two individual components – the *metencephalon* (pons, cerebellum and some cranial nerve nuclei) and the *mylencephalon* (medulla oblongata). Clinically, it is easier to refer to the brainstem, which really is the last part of the brain and is composed of the midbrain + pons + myelencephalon (medulla oblongata). In older texts, the diencephalon is included, but for our purposes, the above description will suffice.

The brainstem is directly continuous with the spinal cord at the ***foramen magnum***. This site is crucially important as it acts as a conduit for information passing to and from the periphery to the central processing unit of the brain. If there is an increase in pressure within the cranial cavity, due e.g., to a *space-occupying lesion* (SOC), *intracranial hemorrhage* or *traumatic brain injury*, it can result in pressure at the foramen magnum. The only way the brain has to move in this enclosed space is downwards through the foramen magnum. This is called *brain herniation*, or *coning*, and results in pressure on the brainstem, and affects the components in that part of the brain, i.e. cardiorespiratory functions. This is a life threatening condition and requires urgent neurosurgical attention.

The brain is comprised, like the spinal cord, of two types of matter – gray and white. The gray matter of the brain is found in the outer aspect of the brain in the cerebral cortex. The gray matter is comprised of neurons, whereas the white matter is composed of supporting glial cells and myelinated axons. The white appearance (especially in tissue which has been fixed in formaldehyde) is due to the lipids of the myelin. The main substance of the brain is white matter but dispersed throughout it are areas of gray matter referred to as the basal ganglia. These subcortical nuclei are designed for a variety of functions including voluntary motor control, emotions, cognition, eye movements and learning. They will be discussed in detail later in the relevant chapters.

1.1.1.2 Spinal Cord

The spinal cord is a long cylinder which occupies the upper two-thirds of the *vertebral canal*. Unlike the brain, the spinal cord gray matter is located within the main substance of it, and is surrounded by the white matter. It is the opposite way round in the brain. On the lateral aspects of the spinal cord, is a pair of *spinal roots*. Each side is composed of a ventral and dorsal root depending on whether it arises from the anterior (ventral) or posterior (dorsal) aspect of the spinal cord.

In summary, there are 31 pairs of spinal roots with their corresponding dorsal and ventral roots. There are eight *cervical*, twelve *thoracic*, five *lumbar*, five *sacral* and one *coccygeal*. Each of these combinations of dorsal and ventral roots joins to form a single spinal nerve, which then divides into a *dorsal* and *ventral ramus*. The specific details of the spinal cord will be dealt with later.

1.1.2 Peripheral Nervous System

The peripheral nervous system is the part of the nervous system that is comprised of the cranial, spinal and peripheral nerves, as well as their sensory and motor nerve endings. In other words, it is the part of the nervous system which is comprised of nerves and ganglia which lie out-with the brain and spinal cord (which comprises the CNS). As this part of the nervous system primarily lies out-with the skull and vertebral column, it is prone to damage from trauma and from toxins. Therefore, the PNS is comprised of the 12 pairs of cranial nerves, and the 31 pairs of spinal nerves, i.e. 43 pairs of nerves in total. The nerves of the PNS can be classified as belonging to either afferent (taking information to the CNS) or efferent (away from the CNS). With spinal nerves, they contain both afferent and efferent information, whereas some cranial nerves like the olfactory and optic nerves contain only afferent information (for smell and sight, respectively).

Broadly speaking, there are two main divisions of the PNS – the somatic and the autonomic nervous systems. The somatic nervous system terminates on the skeletal muscle, whereas the autonomic nervous system supplies all structures other than the skeletal muscle, e.g. glands and smooth muscle.

1.1.2.1 Neurons

A neuron is a highly specialized cell which is capable of communicating (rapidly) via electrical signals and has the presence of a synapse – a specialized area to allow the passage of electrical activity. These synapses are in two forms:

(1) *Chemical synapse* – electrical activity in one neuron passes to the presynaptic terminal. Here, a neurotransmitter is released and passes to the postsynaptic cell (dendrite) to takes its effect via receptors in the postsynaptic density.

(2) *Electrical synapse* – the pre and postsynaptic cells are connected via gap junctions. These regions are specialized in passing electrical current from one neuron to the next.

A neuron has some key anatomical features which are summarized in the table below (Table 1.1).

In the central nervous system, several different types of neuron are found. The most common type of neuron found in the central nervous system are called *multipolar neurons*, due to having many types of dendrites, as well as the single axon. The following provides a summary of the main types of neurons.

- *Multipolar neurons*. These have at least 2 dendrites which extend from the neuronal soma. Multipolar neurons are also classified as Golgi Type 1 neurons, and Golgi Type 2 neurons. Golgi Type 1 neurons have long axons which originate in the gray matter of the spinal cord. Golgi Type 1 neurons are typically found in the ventral gray horn of the spinal cord. They are also typical of *pyramidal neurons* of the cerebral cortex or *Purkinje* neurons of the cerebellum. On the other hand, Golgi Type 2 neurons either do not have an axon at all, or if they do, the axon does not exit from the gray matter of the central nervous system. Golgi Type 2 neurons are found in the granular layer of the cerebellum and hippocampus. These neurons are called *granule cells*, typically found throughout the cerebral cortex, cerebellum, hippocampus, olfactory bulb and the dorsal cochlear nucleus.

- *Bipolar neurons*. These neurons are less typical and have an axon and dendrite (or extension of the axon) at opposite sides of the neuronal cell body. These neurons are typically involved in

Table 1.1. The Key Anatomical Features Common to a Neuron, and Details the Main Functions of Each Component

Anatomical Feature of Neuron	Function
Soma (cell body)	Protein synthesis (abundance of Nissl substance/body) Location of neurofilaments (maintenance of neuron and structural support) "Powerhouse" of neuron Location of nucleus, nucleolus and Nissl body (location of rough endoplasmic reticulum (and ribosomes)
Dendrites	Cellular extensions Majority of input to neuron arrives here via dendrites (via the dendritic spine)
Axon	Transmission of the electrical impulse (action potential) away from the neuronal cell body (soma) To allow for communication with nearby neurons
Axon terminal	Dilated terminal region of the axon Release of neurotransmitter (from the vesicles) into the synaptic cleft to communicate with the dendrite of the next neuron it targets
Axon hillock	The region close to the soma where the axon originates from Location of the voltage gated sodium channels Most excitable part of the neuron May receive information into this point too
Myelin sheath	Propagation of electrical impulses along the axon Increased electrical resistance No voltage gated channels
Nodes of Ranvier	Location of voltage gated ion channels Location of ion exchangers (e.g. Na^+/K^+ and Na^+/Ca^{2+}) Aids rapid propagation of electrical impulses along the axon
Synapse	Structure capable of transmitting chemical or electrical signals Position of pre and postsynaptic area for communication between two neurons (e.g. axon terminal of one neuron and dendrite of adjacent neuron)

transmission of information related to the special senses. They are found in the retina (transmission of visual information), olfactory epithelium (transmission of information related to smell) and the vestibulocochlear nerve (transmitting information related to sound and balance).

- *Unipolar neurons.* These are also referred to as *pseudounipolar* neurons. They have a single axon which extends both centrally and peripherally. The central portion of this neuron extends into the spinal cord and the peripheral portion will extend into the periphery, terminating perhaps in the skin, muscle or joints. *Pseudounipolar* neurons do not have dendrites and are typically found in the *dorsal root ganglia.*

- *Anaxonic neurons.* These are neurons where no obvious axon is identifiable from the dendritic tree and are typically found within the retina and the brain.
- *Betz neurons.* These neurons are the largest of all neurons and are found within the primary motor cortex (of the frontal lobe). These neurons send their axons through the corticospinal (discussed in Chapter 9) tract to reach the ventral horn cells, which contains the motor neurons.

Another classification that exists for neurons is interneurons. These are neurons that connect two neurons together and can either be motor or sensory, but can also be classified as excitatory or inhibitory (more common in the central nervous system). Interneurons can also be thought of as local circuit neurons. Examples of interneurons are now given below.

- *Spindle cells.* These are defined based on their appearance. They have also been referred to as Von Economo neurons. Their soma is spindle shaped in appearance and have a single axon at the apex of the cell with a single dendrite running in the opposite direction. Spindle cells are found in the fronto-insular cortex and anterior cingulate cortex. More recently, they have also been found in the dorsolateral prefrontal cortex (Fajardo et al., 2008).
- *Lugaro cells.* Inhibitory sensory interneurons located within the cerebellum. These neurons interconnect a large number of neurons within the cerebellum.
- *Basket cells.* These neurons are inhibitory interneurons located within the cerebral cortex as well as the hippocampus and the cerebellum. These neurons use the inhibitory transmitter GABA as their way to communicate with other neurons. They have also been subclassified into three different subtypes – large, small and nest basket cells based on their appearance (Wang et al., 2002).
- *Unipolar brush cells.* These neurons have a single dendrite but at the end point of this, many short "brush-like" structures arise from it, thus increasing the surface area for them to communicate with. These neurons are excitatory and use the excitatory transmitter glutamate. They are typically found within the cerebellar cortex (granular layer)

It must also be noted that in the peripheral nervous system, neurons are found in ganglia, which is a collection of nerve cell bodies. These are found either in layers, or laminae, or in groups called nuclei in the central nervous system.

1.2 DIVISIONS OF THE NERVOUS SYSTEM

Broadly speaking, the nervous system can be thought of as being divided into its structural and functional components.

1.2.1 Structural Division of the Nervous System

1.2.1.1 Central Nervous System

Structurally, the nervous system can be classified as the CNS and the PNS, as previously mentioned. The CNS is comprised of the *brain* and *spinal cord*, and this will now be dealt with in a little more detail. The brain is primarily composed of three main areas which develop embryologically – the forebrain (prosencephalon), midbrain (mesencephalon) and hindbrain (rhombencephalon). During development in mammals, the forebrain continues to grow, whereas in other vertebrates, e.g. amphibians and fish, the three divisions remain in proportion to each other during growth.

The brain can also be subdivided into the following.

(1) *Telencephalon* (cerebral hemispheres) + *Diencephalon* (thalamus and hypothalamus) = FOREBRAIN
(2) *Mesencephalon* = MIDBRAIN
(3) *Metencephalon* (pons, cerebellum and the trigeminal, abducent, facial and vestibulocochlear nerves) + *Myelencephalon* (medulla oblongata plus the glossopharyngeal, vagus, accessory and hypoglossal nerve nuclei) = HINDBRAIN

1.3 BRAIN

1.3.1 Forebrain

Surrounding the core of the forebrain, i.e. the diencephalon, is the two large cerebral hemispheres (left and right), which constitutes the cerebrum. The cerebrum is composed of three regions.

(1) Cerebral cortex

The cerebral cortex is the gray matter of the cerebrum. It is comprised of three parts based on its functions – motor, sensory and association areas. The motor area is present in both cerebral cortices. Each one controls the opposite side of the body, i.e. the left motor area controls the right side of the body, and vice versa. There are two broad regions – a primary motor area responsible for execution of voluntary movements, and supplementary areas involved in selection of voluntary movements.

The sensory area receives information from the opposite side of the body, i.e. the right cerebral cortex receives sensory information from the left side of the body. In essence it deals with auditory information (via the primary auditory cortex), visual information (via the primary visual cortex) and sensory information (via the primary somatosensory cortex).

The association areas allow us to understand the external environment. All of the cerebral cortex is subdivided into lobes of the brain. These are as follows.

(a) Frontal lobes

Broadly speaking the frontal lobe deals with "executive" functions and our long-term memory. It also is the site of our primary motor cortex, toward its posterior part.

(b) Parietal lobes

The parietal lobes are responsible for integration of sensory functions. It is the site of our primary somatosensory cortex.

(c) Temporal lobes

The temporal lobes integrate information related to hearing, and therefore, is the site of our primary auditory cortex.

(d) Occipital lobes

The occipital lobes integrate our visual information and functions as the primary visual cortex.

(2) Basal ganglia

The basal ganglia are three sets of nuclei – the *globus pallidus*, *striatum* and *subthalamic nucleus*. These nuclei are found at the lower end of the forebrain and are responsible for voluntary movement, development of our habits, eye movements and our emotional and cognitive functions.

(3) Limbic system
The limbic system is comprised of a variety of structures on either side of the thalamus. It serves a variety of functions including long-term memory, processing of the special sense of smell (olfaction), behavior and our emotions.

1.3.1.1 Thalamus

The thalamus is like a junction point of information. It is a relay point for all sensory information (apart from that related to smell). It also functions in the regulation of our wakened state, or sleep. In addition, it provides a connection point for motor information on its way to the cerebellum.

1.3.1.2 Hypothalamus

The hypothalamus, as its name suggests, is located below the thalamus. It secretes hormones influencing the pituitary gland, and in turn, a wide variety of bodily functions. It regulates autonomic activity ranging from temperature control, hunger and our circadian rhythm and thirst.

1.3.2 Midbrain

The midbrain, as its name suggests, is found between the hindbrain below and the cerebral cortices above. Comprised of the *cerebral peduncles*, *cerebral aqueduct* and the *tegmentum*, it is involved in motor function, arousal state, temperature control, and visual and hearing pathways.

1.3.3 Hindbrain

The lowest part of the brain developmentally is the hindbrain and comprises the pons, medulla and the cerebellum. These areas control movement, cardiorespiratory functions and a variety of bodily functions like hearing and balance, facial movement, swallowing and bladder control. Therefore brainstem death, i.e. death of these regions, is incompatible with life.

1.4 SPINAL CORD

In the gray matter of the spinal cord, three broad categories exist: (a) sensory cells concerned with sensory and reflex arcs; (b) motor cells leaving by the ventral roots to supply skeletal muscle and (c) motor cells leaving by the ventral roots to go on to supply the autonomic ganglia. On examining the spinal cord in cross-section, there is a butterfly shaped gray matter surrounded by white matter (discussed later). The cells found in the gray matter are composed of cell columns in the rostro-caudal axis. Here there are cell bodies, axons and dendrites, both of the myelinated and unmyelinated types.

In each half of the spinal cord there are three funiculi: the dorsal funiculus (between the dorsal horn and the dorsal median septum), the lateral funiculus (located where the dorsal roots enter and the ventral roots exit) and the ventral funiculus (found between the ventral median fissure and the exit point of the ventral roots).

Based on detailed studies of neuronal soma size (revealed using the Nissl stain), Rexed (1952) proposed that the spinal gray matter is arranged in the dorso-ventral axis into laminae and designated them into ten groupings of neurons identified as I – X (Figure 1.1).

Lamina I contains the terminals of fine myelinated and unmyelinated dorsal root fibers that pass first through the zone of Lissauer

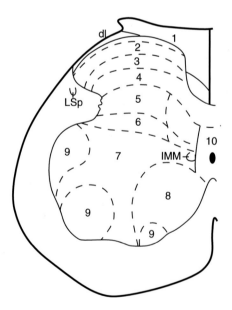

Fig. 1.1. Cross-section of the right side of the spinal cord showing the position of Rexed's laminae. 1–10 indicate the position of laminae I–X, respectively. dl – dorsolateral funiculus; IMM – intermediomedial nucleus; LSp – lateral spinal nucleus.

(dorsolateral funiculus) and then enter lamina I mediating pain and temperature sensation (Christensen and Perl, 1970; Menétrey et al., 1977; Craig and Kniffki, 1985; Bester et al., 2000). The neurons here have been divided into small neurons and large marginal cells characterized by wide-ranging horizontal dendrites (Willis and Coggeshall, 1991). They then synapse on the posteromarginal nucleus. From here the axons of these cells pass to the opposite side and ascend as the lateral spinothalamic tract.

Lamina II is immediately below lamina I, referred to as the substantia gelatinosa. Neurons here modulate the activity of pain and temperature afferent fibers. This lamina has been subdivided into an outer (dorsal) lamina II (II_O) and an inner (ventral) lamina II (II_i) based on the morphology of these layers with stalked cells found in larger numbers in lamina II_O but stalked and islet cells were found throughout lamina II (Todd and Lewis, 1986). Lamina II is the region which receives an extensive unmyelinated primary afferent input, with very little from large myelinated primary afferents (except for distal parts of hair follicle afferents in some animals; Willis and Coggeshall, 1991). The axonal projections from here are wide and varied with some neurons projecting from the spinal cord (projection neurons), some passing to different laminae and some with axons confined to a lamina in the region of the dendritic tree of that cell, e.g. intralaminar interneurons, local interneurons and Golgi Type II cells (Todd, 1996).

Lamina III is distinguished from lamina II in that it has slightly larger cells, but with a neuropil similar to that of lamina II. The classical input to this lamina comes from hair follicles and other types of coarse primary afferent fibers which includes Pacinian corpuscles and rapidly and slowly adapted fibers.

Lamina IV is a relatively thick layer that extends across the dorsal horn. Its medial border is the white matter of the dorsal column, and its lateral border is the ventral bend of laminae I–III. The neurons in this layer are of various sizes ranging from small to large and the afferent input here is from collaterals and from large primary afferent fibers (Willis and Coggeshall, 1991). Input also arises from the substantia gelatinosa (lamina II) and contributes to pain, temperature and crude touch via the spinothalamic tract (Siegel and Sapru, 2006).

Lamina V extends as a thick band across the narrowest part of the dorsal horn. It occupies the zone often called the neck of the dorsal horn. It has a well-demarcated edge against the dorsal funiculus, but an indistinct lateral boundary against the white matter due to the many longitudinally oriented myelinated fibers coursing through this area. The cell types are very homogenous in this area, with some being slightly larger than in lamina IV (Willis and Coggeshall, 1991). Again, like lamina IV, primary afferent input into this region is from large primary afferent collaterals as well as receiving descending fibers from the corticospinal and rubrospinal tracts with axons also contributing to the spinothalamic tracts (Siegel and Sapru, 2006). In addition, in the thoracolumbar segments (T1–L2/3), the reticulated division of lamina V contains projections to sympathetic preganglionic neurons (Cabot et al., 1994).

Lamina VI is present only in the cervical and lumbar segments. Its medial segment receives joint and muscle spindle afferents, with the lateral segment receiving the rubrospinal and corticospinal pathways. The neurons here are involved in the integration of somatic motor processes.

Lamina VII present in the intermediate region of the spinal gray matter contains Clarke's nucleus extending from C8–L2. This nucleus receives tendon and muscle afferents with the axons of Clarke's nucleus forming the dorsal spinocerebellar tract relaying information to the ipsilateral cerebellum (Snyder et al., 1978). Also within lamina VII are the sympathetic preganglionic neurons constituting the intermediolateral cell column in the thoracolumbar (T1–L2/3) and the parasympathetic neurons located in the lateral aspect of the sacral cord (S2–4). In addition Renshaw cells are located in lamina VII and are inhibitory interneurons which synapse on the alpha motor neurons and receive excitatory collaterals from the same neurons (Renshaw, 1946; Siegel and Sapru, 2006).

Lamina VIII and IX are found in the ventral gray matter of the spinal cord. Neurons here receive descending motor tracts from the cerebral cortex and the brainstem and have both alpha and gamma motor neurons, which innervate skeletal muscles (Afifi and Bergman, 2005). Somatotopic organization is present where those neurons innervating the extensor muscles are ventral to those innervating the flexors, and neurons innervating the axial musculature are medial to those innervating muscles in the distal extremities (Siegel and Sapru, 2006).

Lamina X is the gray matter surrounding the central canal and represents an important region for the convergence of somatic and visceral primary afferent input conveying nociceptive and mechanoreceptive information (Nahin et al., 1983; Honda, 1985; Honda and Lee, 1985; Honda and Perl, 1985). In addition, lamina X in the lumbar region also contains preganglionic autonomic neurons as well as an important spinothalamic pathway (Ju et al., 1987a,b; Nicholas et al., 1999).

The *white matter* of the spinal cord contains the ascending and descending pathways. Some of these pathways ascend and descend to and from the brain, whereas others will connect to various levels within the spinal cord itself.

1.4.1 Peripheral Nervous System

The nerves in the peripheral nervous system (PNS) transmit information from all parts of the body to and from the CNS. In total, there are 43 pairs of nerves in the PNS – 12 cranial nerves and 31 spinal nerves. The nerves of the PNS can either be myelinated (formed by the surrounding Schwann cells) or unmyelinated in nature.

During the latter stage of fetal development and into the postnatal period, myelin sheaths are laid down around the axons. It is laid down in successive layers around the axon, from close to its origin all the way to its final branching at its effector tissue. Myelin is an insulating layer around the axon aiding the transmission of electrical impulses. Within the central nervous system, myelin is produced by oligodendrocytes. In the peripheral nervous system, myelin is produced by Schwann cells. Myelin is composed of a variety of different substances but the majority of it is made from lipids. Other significant components of myelin include protein, cholesterol, *Cerebroside* (monoglycoslceramide) and *galactolipid* (glycolipid with its sugar as galactose). Other constituents of myelin, in humans, include *ethanolamine phosphatide, lecithin, sulfatide, plasmogens* (ether phospholipid), *sphingomyelin, phosphatidylserine* and *phosphatidylinositiol* (Siegel et al., 1999). In situ, myelin has water content of approximately 40% but the dry mass of both the peripheral and central nervous system myelin content is primarily made of lipids (approximately three-quarters; Siegel et al., 1999).

A nerve fiber is classified as the axon of the neuron together with its associated myelin sheath and any supporting, or glial cells. The larger the axon, generally, the larger the myelin sheath. Of the entire external diameter of the fiber, the axon comprises approximately two-thirds of that diameter (Kiernan and Rajakumar, 2014). In the peripheral nervous system a nerve is a collection of nerve fibers which are visible without the aid of a microscope. These are bundles of many, many axons, perhaps also with their ensheathing myelin. Within the central nervous system these would be classified as tracts.

The entire nerve is surrounded by dense connective tissue called its *epineurium*. Fibers from this connective tissue pass inwards enclosing nerve fibers called *fasciculi*. The connective tissue that surrounds each fascicle is termed the *perineurium*. Enclosing the individual nerve fibers is yet more connective tissue, and this is referred to as its *endoneurium*.

Nerve fibers can be classified according to the structures that they supply, i.e. related to its functions. Fibers that carry information from the periphery in a sensory ending are referred to as a sensory, or afferent, fiber. Nerve fibers which reach muscle (skeletal) are called motor, or efferent fibers. Indeed, fibers which are targeting smooth muscle and glands are also motor fibers, and sensory fibers also carry information related to the visceral organs. Therefore, a further classification system to allow for differentiation of these features has been accepted. The details of this further subclassification is shown below, and explained in more detail in relation to function, in the next section.

(1) Efferent
   (a) Special visceral efferent – motor fibers which target skeletal muscle. These fibers target muscles of pharyngeal origin. These fibers are specifically carried in the trigeminal, facial, glossopharyngeal, vagus and accessory cranial nerves.
   (b) Somatic efferent – these are motor fibers which target skeletal muscle.
   (c) General visceral efferent – these fibers are motor fibers which are of autonomic in origin, e.g. cardiac, glandular and smooth muscle

(2) Afferent
   (a) General somatic afferent – these fibers are sensory fibers that carry information related to general sensation, e.g. touch, pain and temperature.
   (b) General visceral afferent – these fibers carry sensory information related to the visceral organs. They arise from the glands, blood vessels and the viscera specifically. They are generally classified as those belonging to the autonomic nervous system. The cranial nerves that carry these impulses are the facial, glossopharyngeal and vagus nerves.
   (c) Special visceral afferent – these fibers carry information related to pharyngeal arch origin related to the gastrointestinal tract. They carry information related to taste and smell. These fibers are carried in cranial nerves, specifically the olfactory, facial, glossopharyngeal and vagus nerves.

1.4.2 Spinal Nerves

There are 31 pairs of spinal roots with their corresponding dorsal and ventral roots. There are eight *cervical*, 12 *thoracic*, five *lumbar*, five *sacral* and one *coccygeal*. Within the spinal roots, there are dorsal and ventral roots. In the ventral roots, there are motor fibers, and it is those fibers that supply the skeletal muscle. Within the ventral roots of the thoracic, upper lumbar and some of the sacral levels autonomic fibers are also present. Within the dorsal roots, there are sensory fibers from the skin, subcutaneous and deep tissue, and frequently from the viscera too. The spinal nerve is formed by both the dorsal and ventral root and contains most of the fiber components found in that root. A major peripheral nerve therefore contains the sensory, motor and autonomic fibers within it. Smaller branches however will vary in their composition. Therefore, a nerve to skin will lack motor fibers to the skeletal muscle, but it will contain sensory fibers and autonomic fibers to the vasculature and may also contain fibers supplying the autonomic innervation to the hair follicles.

The dorsal rami will transmit information from the muscles of the back and also the skin. The ventral rami innervate the rest of the trunk and also the limbs. The ventral rami that supply the thorax and abdomen remain relatively separate in their course. However, in the cervical or lumbar and sacral regions, the ventral rami are found intertwined in what is called a plexus of nerves. It is from these plexuses that the peripheral nerves will then emerge.

A simple way to look at it is that when the ventral rami enter into this plexus, its individual components will contribute to several of the peripheral nerves. Indeed, each peripheral nerve therefore will contain fibers from one or more spinal nerves.

Each spinal nerve has a distribution called a *dermatome*. This is the area that is supplied by a single spinal nerve by its sensory fibers running in its dorsal root. However, sectioning of a single spinal nerve rarely will result in complete loss of sensation, or anesthesia. This is because other adjacent spinal nerves will also be carrying fibers from that site. Therefore, the more likely scenario is a reduced level of sensation, or hypoesthesia.

There are many versions of *dermatome maps* which can be used clinically to demonstrate the site(s) of pathology of a patient with a suspected spinal nerve lesion. The most common one to be used clinically is the American Spinal Injury Association's (ASIA) worksheet produced as the International Standards for Neurological Classification of Spinal Cord Injury (ISNCSCI, 2014). This is discussed in considerable detail in Chapter 8.

For motor fibers, carried in the ventral root, they tend to supply more than one muscle. Therefore, each muscle will receive innervation from more than one spinal nerve. Sectioning of a single spinal nerve will result in weakness of more than one muscle. Sectioning of a peripheral nerve will however result in paralysis of that single muscle. The group of muscles that a single spinal nerve supplies is called a *myotome*. Testing of the myotomes is something routinely carried out in the neurological examination of a patient, directed as to the signs and symptoms that the person presents with? Again, this is discussed in more detail in Chapter 8.

1.4.3 Functional Division of the Nervous System

The following diagram (Figure 1.2) summarizes the functional divisions of the nervous system. Central to this is the brain and spinal cord (CNS). Information from the periphery arriving into the CNS is referred to as

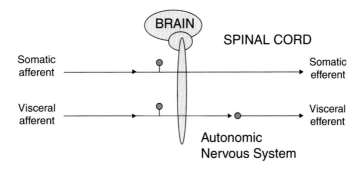

Fig. 1.2. A diagram showing the functional divisions of the nervous system – somatic and visceral. Information arriving into the CNS is referred to as afferent, and information leaving it is efferent. Note the 2 neuron pathway in the autonomic (visceral) nervous system.

afferent. Information exiting the CNS is referred to as efferent. There are two afferent inputs which enter the CNS – somatic and visceral.

TIP!

The easy way to remember what information arrives into and leaves the CNS is rather easy.
*A*FFERENT – *A*RRIVES into the CNS. *A* for *A*FFERENT, *A* for *A*RRIVES!
*E*FFERENT – *E*XITS the CNS. *E* for *E*FFERENT, *E* for *E*XIT!

1.4.4 Somatic Nervous System

The somatic nervous system consists of the cell bodies located in either the brainstem or the spinal cord. They have an extremely long course as they do not synapse after they leave the CNS until they are at their termination in skeletal muscle. They consist of large diameter fibers and are ensheathed with myelin. They are commonly referred to as motor neurons due to their termination in skeletal muscle. Within the muscle fibers, they release the neurotransmitter acetylcholine and are only excitatory, i.e. result only in contraction of the muscle.

1.4.5 Autonomic Nervous System

The visceral, or autonomic nervous system (ANS), can be thought of as that part of the nervous system supplying all other structures apart from skeletal muscle (supplied by the somatic nervous system). However, part of the ANS supplies the gastrointestinal system and is referred to as the enteric nervous system as the neurons are found supplying the glands and smooth muscle in the actual wall of the tract.

Within the ANS generally, it is composed of two neurons and a synapse (as shown in Figure 1.2). This is different to the single neuron of the somatic nervous system. The origin of the first neuron of the ANS is found in the CNS, with the first synapse occurring in an autonomic ganglion. This part is defined as the preganglionic fiber. After the synapse in the autonomic ganglion, the second fiber is referred to as the postganglionic fiber as it passes to the effector organ, in this case cardiac or smooth muscle, glands or gastrointestinal neurons.

The ANS is subdivided into sympathetic and parasympathetic divisions based on physiological and anatomical differences. The sympathetic division arises from the thoracolumbar region from the first thoracic to the second lumbar level (T1–L2). The parasympathetic division arises from cranial and sacral origins. Specifically, the parasympathetic division arises from four of the cranial nerves – the oculomotor (III), facial (VII), glossopharyngeal (IX) and vagus (X) nerves. It also arises from the sacral plexus at the levels of the second to fourth sacral segments (S2–4).

1.4.6 Sympathetic and Parasympathetic Nervous System
The sympathetic nervous system is part of the nervous system that deals with "fight or flight" responses. The parasympathetic nervous system can be classified as part of the nervous system that controls "rest and digest". A summary table is given below comparing what functions each part of the autonomic nervous system causes to a variety of areas around the body (Table 1.2).

1.5 CRANIAL NERVES

There are 12 pairs of cranial nerves which are associated with the brain. These nerves are unique in the human body and may carry one, some or many different types of fibers within them. Cranial nerves are very different to spinal nerves (or roots) in that they do not contain dorsal and ventral roots. Indeed, some cranial nerves may have one ganglion, several, or none at all. Indeed, the optic nerve, responsible for conveying

Table 1.2. Comparison of the Differences at Various Regions of the Body of the Sympathetic and Parasympathetic Nervous System

	Sympathetic	Parasympathetic
Heart	Increases heart rate	Reduces heart rate
	Increases contractility of atria and ventricles	Reduces contractility of atria and ventricles
	Increases conduction	Reduces conduction
Lungs	Relaxes bronchial muscle	Contracts bronchial muscle
	Reduced secretions (via $\alpha 1$ receptors)	Stimulates secretions (via $\alpha 1$ receptors)
Stomach and intestines	Reduced tone and motility	Increased tone and motility
	Contracts sphincters	Relaxes sphincters
	Inhibits secretions	Stimulates secretions
Pancreas	Inhibits exocrine secretion	Stimulates secretion
	Inhibits insulin secretion	Stimulates insulin secretion
Eyes	Contracts radial muscle (dilates pupil)	Contracts sphincter muscle (constricts pupil)
	Relaxes ciliary muscle (for far vision)	Contracts ciliary muscle (for near vision)
Nasal, lacrimal and salivary glands	No significant effect	Stimulation of serous and mucous secretions from the secretory cells
Skin	Contracts arrector pili muscles (hair to stand on end)	N/A
	Localized secretion of sweat glands	Generalized secretion of sweat glands
Urinary bladder	Relaxes wall	Contracts wall
	Contracts sphincter	Relaxes sphincter
Genital organs	May stimulate vasoconstriction, but uncertain and variable	May stimulate glands and smooth muscle; vascular dilatation
Adrenal gland	Stimulation of secretory cells to produce epinephrine	No effect
Arterioles	Variable	Dilates coronary and salivary gland arterioles (via $\alpha 1,2$ receptors)

The different divisions of the autonomic nervous system affect each territory in very contrasting ways

the sensation for vision, is deemed a direct outgrowth of the brain, and can be compared to as a fiber tract from the central nervous system.

Cranial nerves arise from the brain, as distinct from spinal nerves which arise from the spinal cord. There are 12 pairs of cranial nerves. Some are purely motor (e.g. hypoglossal (XII) which supplies the tongue muscles); some are purely sensory (e.g. the optic (II) nerves which come from the retina); some are mixed (e.g. the trigeminal (V) nerve which is sensory to the face and scalp and motor to the muscles of mastication). We can consider

(1) where they are attached to the brain
   (a) forebrain-olfactory (I) and optic (II) nerves
   (b) midbrain-oculomotor (III) and trochlear (IV) nerves. (The trochlear is unique in arising from the dorsal surface of the brainstem); and
   (c) hindbrain-trigeminal (V) from the pons ; abducens (VI), facial (VII) and vestibulocochlear (VIII) from the pontomedullary junction ; glossopharyngeal (IX) vagus (X) accessory (XI) and hypoglossal (XII) from the medulla
(2) how they are related to embryological development.

Many of the structures in the head and neck arise from two quite distinct embryological sources. These sources are

(1) *Somites*. These are paired segmental blocks of tissue which run along the length of the embryo, rather like a series of building bricks. They give rise to many structures, including muscles. Motor cranial nerves which supply somite-derived muscles are called somatic efferent* nerves and consist of cranial nerves III, IV, VI which supply the extraocular muscles (which move the eyes about) and cranial nerve XII to the muscles of the tongue.
(2) *Branchial arches*. During development, the embryo passes through a stage of having pharyngeal or branchial arches at the side of the neck – exactly as a fish has gill arches. These arches form a numbered series and, again, give rise to many adult structures, including muscles. Motor cranial nerves which supply them are termed branchial efferent* nerves. They are (Table 1.3).

Table 1.3. The Cranial Nerves Associated With Each of the Branchial Arches, and the Muscles They Supply

Arch		Nerve	Muscles
I	Mandibular arch	Trigeminal (V)	Mastication
II	Hyoid arch	Facial (VII)	Facial expression
III–VI		IX, X, XI	Laryngeal/pharyngeal

* "Efferent" means "going away from". In the context of cranial or spinal nerves, efferent nerves are those going out from the CNS, i.e. motor nerves. The opposite is "afferent" (=going toward) which would describe sensory nerves taking information to the central nervous system.

Here, each cranial nerve will be dealt with briefly, in terms of the ganglion (ganglia) associated with it, the type of fibers found within, and the function of each one.

1.5.1 Olfactory Nerve

The first cranial nerve is the *olfactory nerve*. It is responsible for conveying *special sensory* fibers in relation to smell. The cell bodies of this nerve are found in the olfactory organ in the upper part of the *nasal cavity*, *nasal septum* and the *superior concha*. When an odor activates the cilia, electrical impulses then passes to the *olfactory receptor cells*, then to the *olfactory bulb* as the olfactory nerve enters the *cribriform plate of the ethmoid bone*, and onwards to a variety of brain regions for interpretation of smell. Specifically, the brain regions responsible for interpretation of smell are the *piriform cortex of the anterior temporal lobe*, *anterior olfactory nucleus*, *amygdala* and *entorhinal cortex*.

1.5.1.1 Typical Pathologies to Affect the Olfactory Nerve

A variety of pathologies may affect this nerve resulting in altered smell, e.g. hypo-osmia, hyperosmia, anosmia or dysosmia. The most common pathology to affect the olfactory nerve is the common cold, resulting in a reduced sense of smell, i.e. hypo-osmia. Conditions associated with a dysfunction of the olfactory nerve can include, but are not limited to the following:

(1) Local factors, e.g. allergies, polyps, tumors or exposure to irritant chemicals
(2) Neurological pathologies, e.g. epilepsy, Parkinson's disease, multiple sclerosis or Alzheimer's disease
(3) Neoplastic. These new growths can occur intranasally or intracranially affecting the course and distribution of the olfactory nerve.
(4) Endocrine, e.g. diabetes mellitus, Cushing's syndrome or insufficiency of the adrenocortical system
(5) Nutritional deficiencies. A variety of insufficient supply of the nutrients can affect the sense of smell and can include vitamin B12 deficiency.

1.5.1.2 Clinical Testing of the Olfactory Nerve

Testing of this nerve is often missed out on clinical examination, but can include identifying the patency of each nostril first. Then, exposing each nostril separately, each nostril (with the other one occluded) can be tested with gentle, non-noxious stimuli, e.g. vanilla essence or peppermint oils. Further details on related pathologies and clinical examination of the olfactory nerve is covered in Clinical Anatomy of the Cranial Nerves (Rea, 2014), the companion text to this book.

1.5.2 Optic Nerve

The second cranial nerve is the optic nerve. It is an unusual cranial nerve in that it develops from the *diencephalon*. As such, it contains exactly the same layers that surround the brain – the meninges, and also the *cerebrospinal fluid (CSF)* that is contained in the *subarachnoid space*. The optic nerve only conveys special sensory information for vision. Visual information passes to the posterior aspect of the eye reaching the *retina*. The photoreceptor cells then transmit this information to the bipolar cells. These bipolar cells have impulses coming from both the rods (for low light intensity) and cones (for color vision), and then transmit impulses onwards to the ganglion cells and then to the optic nerve.

The optic nerve then enters the cranial cavity at the optic canal and then onwards to the optic chiasm. From this point, nerve fibers on the medial, or nasal side of each retina cross the midline to the optic tract of the opposite side. However, the nerve fibers on the lateral, or temporal side, pass posteriorly but do not cross.

1.5.2.1 Typical Pathologies to Affect the Optic Nerve

(1) *Refractive errors*. These are defined as myopia (short sightedness) and hypermetropia (long sightedness). These defects are due to the light either focusing in front of the retina, or behind it, respectively.

(2) *Visual field defects*. A variety of visual field defects can affect the course of the visual pathways from the retina to the visual cortices, and this depends on where exactly in the path the lesion is. Generally speaking, a lesion anterior to the optic chiasm will result in ipsilateral loss of vision. A lesion at the optic chiasm will result in a bitemporal hemianopia. Lesions posterior to the optic chiasm will result in loss of vision in the field opposite to the side of the damage.

(3) *Papilloedema*. This refers to swelling of the optic disc. It is caused by a variety of conditions, but any presentation of this condition MUST be investigated to identify the cause, and establish the reason, perhaps for raised intracranial pressure.

(4) *Red eye*. Like papilloedema, this condition needs urgent attention to ensure preservation of vision, and to identify and treat the cause.

(5) *Optic neuritis*. This is a condition where there is inflammation of the optic nerve. Again, this condition can result in complete or partial loss of vision and needs urgent attention. It would be wise for an ophthalmologist to be consulted with this condition (and any other condition that may merit it).

1.5.2.2 Clinical Testing of the Optic Nerve

A number of tests should be used to test the integrity of this nerve including visual acuity, visual fields, pupillary size, color assessment and ophthalmoscopy. Clinical examination of this nerve is covered in Clinical Anatomy of the Cranial Nerves (Rea, 2014), the companion text to this book. Many pathologies can affect the optic nerve, and can potentially put vision at risk. These conditions must be fully assessed and evaluated and specialist advice may need to be sought immediately.

1.5.3 Oculomotor Nerve

The third cranial nerve is the *oculomotor nerve*. It contains somatic motor and visceal motor fibers. It arises from the midbrain and gains entry into the orbit via the superior orbital fissure. It supplies the majority of the extraocular muscles, but also carries with it parasympathetic innervation to innervate the sphincter and ciliary muscles. The sphincter muscle will be activated by the parasympathetic nervous system to result in pupillary constriction. The ciliary muscle will be activated by the parasympathetic nerve (via the ciliary ganglion) resulting in the lens increasing its convexity for near vision to be enhanced. Specifically, the oculomotor nerve supplies the superior, medial and inferior recti and the inferior oblique. The oculomotor nerve also supplies the levator palpebrae superioris, which is responsible for maintaining the eye open.

1.5.3.1 Typical Pathologies to Affect the Oculomotor Nerve

There are many conditions which can affect the oculomotor nerve. These range from vascular causes (aneurysms of the posterior communicating, posterior cerebral or superior cerebellar arteries, or even cavernous sinus thrombosis), neurological pathologies (space-occupying lesions), autoimmune disorders, infections or trauma to the base of skull. As the oculomotor nerve supplies so many of the extraocular muscles of the eyeball, the patient would have a ptosis and unopposed "down and out" appearance of the affected eye. This is due to the pull of the intact superior oblique and lateral rectus muscles (supplied by the trochlear and abducent nerves, respectively). Ptosis results from the lack of innervation of the muscle which retains the eye "open", i.e. levator palpebrae superioris.

1.5.3.2 Clinical Testing of the Oculomotor Nerve

Clinical examination of this nerve is covered in Clinical Anatomy of the Cranial Nerves (Rea, 2014), the companion text to this book. However, briefly the oculomotor (and indeed the trochlear and abducent) nerves can be tested for first by observing the eye in its resting state. This is to enable identification of any nystagmus. Then, slowly, an imaginary H-shape should be drawn in the air, either by using the examiners finger, or by a pencil. Have the patient keep their head still, and only with their eyes, follow this H shaped pattern being drawn out in space. More advanced testing can be done, dependent on the clinical situation.

1.5.4 Trochlear Nerve

The fourth cranial nerve is the trochlear nerve and contains only somatic motor fibers. It supplies a single extraocular muscle – the superior oblique – a muscle responsible for intorsion of the eye with a small amount of abduction and depression. It arises from the trochlear nucleus at the level of the midbrain. It has the longest course of all cranial nerves, yet is also the smallest one too. As it has such a long intracranial course, this makes it susceptible to any raised intracranial pressure. Typically, pathology of this nerve will result in the patient complaining of diplopia, especially when looking down. This is typically revealed when the patient walks down stairs.

It passes to the superior oblique muscle by exiting the skull at the superior orbital fissure, and is also closely related t the superior cerebellar and posterior cerebral arteries.

1.5.4.1 Typical Pathologies to Affect the Trochlear Nerve

Causes of a trochlear nerve pathology are wide and varied, especially due to its long intracranial course. It may be affected by trauma (e.g. by a road traffic accident causing a fractured base of skull), vascular (hypertension, microvascular disease or aneurysm), neurological (multiple sclerosis or myasthenia gravis), infection, neoplastic (any space-occupying lesion), congenital or iatrogenic pathologies, e.g. from recent neurosurgery.

1.5.4.2 Clinical Testing of the Trochlear Nerve

Clinical testing of the trochlear nerve is the same as for the oculomotor nerve – the H-shape being drawn in space, and observation of the eyeball. More specific testing may have to be done using CT/MRI as dictated by the clinical presentation of the patient. More detail on clinical examination of the trochlear nerve is covered in Clinical Anatomy of the Cranial Nerves (Rea, 2014), the companion text to this book.

1.5.5 Trigeminal Nerve

The fifth cranial nerve is the trigeminal nerve and is the largest of all of the cranial nerves. It contains two major types of fibers within it – general sensory and also branchial motor fibers. As the name suggests, the trigeminal nerve has three branches:

1. Ophthalmic (abbreviated as CN V_1) – general sensory component
2. Maxillary (abbreviated as CN V_2) – general sensory component
3. Mandibular (abbreviated as CN V_2) – general sensory and branchial motor components

The trigeminal nerve is the principal sensory nerve of the head innervating the skin of the face, mucosa of the mouth, nasal cavity and paranasal sinuses, and most of the dura mater and the cerebral arteries.

The trigeminal nerve has two ganglia associated with it – the trigeminal (sensory) and the submandibular (which houses synapses for pre and postsynaptic parasympathetic innervation of the submandibular and sublingual salivary glands from the chorda tympani of the facial nerve). It also has four nuclei related to it: the spinal, pontine and mesencephalic trigeminal nuceli and the trigeminal motor nucleus.

The spinal trigeminal nucleus carries information related to pain, temperature and light touch. The pontine trigeminal nucleus carries information for discriminative and light touch, as well as proprioception of the jaw. The mesencephalic trigeminal nucleus carries proprioceptive information of the lower jaw, and the trigeminal motor nucleus carries motor information from the muscles of mastication, as well as tensor veli palatini, tensor tympani, anterior belly of digastric and mylohyoid.

One of the most clinically relevant features of this nerve, aside from the sensory innervation of the majority of the face, lies in the fact that the inferior alveolar branch (of the mandibular division) is the one dentists routinely anesthesia for lower jaw dental work.

1.5.5.1 Typical Pathologies to Affect the Trigeminal Nerve

A variety of pathologies affect this nerve, including trigeminal neuralgia, herpes zoster and disorders of the temporomandibular joint. These conditions can be extremely debilitating for the patient, and treatment depends on the cause in each of these cases. With trigeminal neuralgia, it can often be caused by a small blood vessel pressing on one of the branches of this nerve. It can be treated conservatively by anticonvulsant drugs, or in more severe cases, microvascular decompression surgery has been proven to have excellent success rates long term (NHS, 2012).

The treatment of herpes zoster has proven to be problematic, with little affect of topical antiviral agents. Oral antiviral agents may reduce the severity and duration of symptoms, but at the moment, there is no cure to this disease.

Temporomandibular joint dysfunction presents in a variety of ways, with a wide variety of causes. Specialist input may be needed from the dental surgeon, or perhaps maxilla-facial experts.

Further explanation of pathologies and clinical examination of the trigeminal nerve is covered in Clinical Anatomy of the Cranial Nerves (Rea, 2014), the companion text to this book. A brief discussion is contained in Chapter 10.

1.5.5.2 Clinical Testing of the Trigeminal Nerve

There are two major components of this nerve that are relevant in clinical testing – its sensory division, and its motor component. Testing of the sensory distribution can be done by using cotton wool over each of the areas it supplies, i.e. the distribution of each of its branches (ophthalmic, maxillary and mandibular). This should be done on both sides of the face, and with the patient's eyes closed to ensure they do not predict when each territory will be touched.

Also, ensure the strength of the supply of the muscles of mastication is assessed. The main ones tested for in clinical examination of the trigeminal nerve are the temporalis and the masseter. This is simply done by asking the patient to clench their teeth, and for the examiner to assess the strength of the muscles over their territories, i.e. temple and angle of the mandible.

Further testing of the trigeminal nerve can be undertaken by assessing the jaw jerk reflex. It may be relevant to assess the trigeminal nerve using nerve conduction studies, or performing an MRI, but specialist advice should be sought based on the patient's history and findings on clinical examination. Further details of the typical pathologies to affect the trigeminal nerve (as well as underpinning anatomy) and clinical examination can be found in the companion text to this book entitled Clinical Anatomy of the Cranial nerves (Rea, 2014). A brief discussion is contained in Chapter 10.

1.5.6 Abducent Nerve

The sixth cranial nerve is the abducent nerve. It is purely a somatic motor nerve and supplies a single extraocular muscle – the lateral rectus muscle. This nerve arises at the lebel of the pons at the facial colliculus. The nerve arises close to the midline and the fibers pass in a ventrocaudal direction close to the corticospinal tract and exits at the pontomedullary junction. It is closely associated with the dura mater and the inferior petrosal sinus. The abducent nerve has the longest intradural course of all cranial nerves, but also passes through the cavernous sinus, along with other cranial nerves (oculomotor, trochlear and ophthalmic and maxillary divisions of the trigeminal nerve).

The abducent nerve then enters the orbit through the superior orbital fissure to enter the lateral rectus on its medial side. As well as containing somatic motor fibers, it can also contain proprioceptive fibers destined for this muscle.

1.5.6.1 Typical Pathologies to Affect the Abducent Nerve

A variety of pathologies may affect the abducent nerve including trauma (e.g. base of skull fractures), vascular (infarction, hemorrhage or aneurysm), neurological (with diabetes mellitus being the most common one), infection (e.g. meningitis), neoplastic (space-occupying lesion intracranially), iatrogenic (perhaps in treatment of cervical spinal injuries with a halo orthosis (Benzel, 2012) and congenital causes.

1.5.6.2 Clinical Testing of the Abducent Nerve

Clinical testing of the trochlear nerve is the same as for the oculomotor nerve – the H-shape being drawn in space, and observation of the eyeball.

There is an easy way to remember what nerves supply the extraocular muscles by the following tip box.

TIP!

The following mnemonic will aid to remembering the supply of the extraocular muscles. It looks like a sulfur compound that we used to encounter in school chemistry.

$$LR_6SO_4$$

It simply means:

Lateral rectus (LR) – supplied by the sixth (6) cranial nerve (abducent nerve)
Superior oblique (SO) – supplied by the fourth (4) cranial nerve (trochlear nerve).

ALL OTHER extraocular muscles are supplied by the oculomotor nerve.

Advanced testing can be undertaken of the abducent nerve, generally by a specialist, e.g. ophthalmologist and/or neurological input. Three broad categories of detailed examination can be done including the rotational movement of each eye, comparison of yoke muscles (pairs of muscles that move the eyes in conjugate direction) and the red lens diplopia test (using diplopia as a test for weakness of the eye muscles) (Walker et al., 1990). Further discussion of these pathologies and clinical examination of the abducent nerve are covered in Clinical Anatomy of the Cranial Nerves (Rea, 2014), the companion text to this book.

1.5.7 Facial Nerve

The seventh cranial nerve is the facial nerve and is an incredibly complex nerve originating from the second pharyngeal arch. It contains four different types of fibers – branchial motor, special sensory, visceral motor and general sensory fibers. Perhaps from the clinical perspective, the branchial (pharyngeal) motor component is most important, as it is that which supplies the muscles of facial expression. Paralysis of one or more of these branches will result in a detrimental affect to the patient resulting in facial paralysis, and perhaps an inability to speak, swallow, close their eyes and convey any form of facial expression.

The facial nerve originates at the pontomedullary junction lateral to the sixth nerve root. The facial nerve (both its motor root and the nervus intermedius (which carries parasympathetic and sensory fibers)) and the vestibulocochlear nerve (eight cranial nerve) pass into the internal auditory meatus of the petrous temporal bone. It then passes close to the middle ear, where it turns posteriorly at the genu (Latin, knee) where the sensory ganglion is also located – the geniculate ganglion. On reaching the posterior aspect of the middle ear, it turns sharply again at the second genu and passes downwards to exit the skull at the stylomastoid foramen.

On exiting the skull, the facial nerve then enters the posterior aspect of the parotid gland. When inside the parotid gland, the facial nerve divides into a superior temporofacial branch and inferior cervicofacial division. These then branch further to provide the terminal branches that supply the muscles of facial expression. The final branches are the temporal, zygomatic, buccal, marginal mandibular and cervical.

The temporal branches go on to supply the frontalis and muscles of the external ear. The zygomatic branches supply the remainder of frontalis, two parts of orbicularis oculi and adjacent muscles. The buccal branches supply the upper half of orbicularis oris, buccinator and dilator muscles inserting into the upper lip. The marginal mandibular branches supply muscles of the lower lip and the cervical branches supply platysma. From the clinical perspective, it is these branches that are most important.

Other branches of the facial nerve which merit discussing here are the intratemporal branches. Within the petrous temporal bone, the following are the major branches relevant also to clinical practice:

- Greater petrosal nerve
  This branch arises from the geniculate ganglion and is joined by the nerve of the pterygoid canal. It contains secretomotor fibers for the lacrimal and nasal glands

- Nerve to stapedius
  Supplies the stapedius muscle, which is responsible for "dampening down" loud noises protecting the middle and inner ear structures
- Chorda tympani nerve
  This nerve joins the lingual nerve (of the mandibular division of the trigeminal nerve – see Chapter 10 on trigeminal nerve) and is distributed to the anterior 2/3 of the tongue. It contains the following:
  - Taste and sensation fibers from the front 2/3 of the tongue and the soft palate
  - Preganglionic secretory and vasodilator fibers that synapse in the submandibular ganglion

In addition to this, there are 3 ganglia which are associated with the facial nerve – the geniculate, pterygopalatine and submandibular ganglia. These are discussed below in relation to their functions:

- Geniculate – Sensory ganglion
  - Special sensory neuronal cell bodies for taste
  - Fibers from motor, sensory and parasympathetic functions pass through here
- Pterygopalatine – synapses here for pre and postsynaptic parasympathetic innervation of the lacrimal glands
  - Sensory and sympathetic fibers also pass through it
- Submandibular – synapses here for pre and postsynaptic parasympathetic innervation of the submandibular and sublingual salivary glands

1.5.7.1 Typical Pathologies to Affect the Facial Nerve

Generally, pathologies of the facial nerve can be classified into either upper or lower motor neuron causes. Upper motor neuron lesions typically involve a stroke, with lower motor neuron causes being wide and varied. The most common cause for lower motor neuron pathology of the facial nerve is due to Bell's palsy. This diagnosis is generally one of exclusion, and can be caused by inflammation of the point at which the facial nerve leaves the skull at the stylomastoid foramen. Other related causes of lower motor neuron pathologies resulting in facial nerve palsy include infections (herpes zoster and herpes simplex), trauma to the temporal bone or direct facial nerve trauma, neurological causes (neuropathy or multiple sclerosis) or neoplastic pathology (typically of the parotid gland).

1.5.7.2 Clinical Testing of the Facial Nerve

Testing of the facial nerve can easily be done at the bedside in the first instance. The facial nerve, from the clinical perspective, is crucial in supplying the muscles of facial expression. The integrity of the facial nerve is tested for by asking the patient to raise their eyebrows; close their eyes against resistance; puff out their cheeks, and by showing their teeth. Further testing can be done regarding the corneal reflex, and also taste from the anterior two-thirds of the tongue. Clinical examination of this nerve, and further explanations of pathologies is covered in Clinical Anatomy of the Cranial Nerves (Rea, 2014), the companion text to this book. A brief discussion is contained in Chapter 10.

1.5.8 Vestibulocochlear Nerve

First, an overview of the apparatus of the ear will be given, followed by a detailed summary of the main nerve for hearing and balance – the vestibulocochlear nerve.

The visible part of the outer ear (the auricle or pinna) is highly variable in size and shape. You should not bother to learn the names of different parts. The pinna is a complex sound-collecting device which directs sound waves toward the external auditory meatus, the passage that ends at the eardrum or tympanic membrane. You can artificially enlarge the pinna by putting your cupped hand behind it. This should increase the volume of sound which you hear. You can even try this yourself! The pinna is flexible because of its elastic cartilage core.

The external auditory meatus is lined by ordinary skin (with hairs at the entrance to it) and possesses ceruminous glands which produce wax. Excessive wax can reduce hearing. The glands are controlled by steroid hormones and are usually larger and more productive in men. The outer part of the meatus is surrounded by cartilage, but near the tympanic membrane it enters the bone of the skull.

The tympanic membrane is a very thin sandwich being made of (i) an outermost layer of skin, (ii) an innermost layer of mucous membrane, and (iii) a middle layer of fibrous tissue. Most of the membrane is taut, but there is a looser flaccid part at the top of the membrane.

Beyond this is the middle ear (tympanic) cavity which lies within the petrous temporal bone. The cavity is linked anteriorly to the nasopharynx via the Eustachian tube, and posteriorly to the air cells of the mastoid process by a small entrance. Conceptually, the eustachian tube/middle ear/mastoid can be seen as a very extended paranasal sinus. It will come as no surprise, therefore, to learn that it is lined by respiratory mucous membrane and that upper respiratory tract infections often affect the middle ear – especially in children where the Eustachian tube is shorter and straighter. This may result in a middle ear which is blocked by mucus ("glue ear") and sometimes the tympanic membrane perforates as a result of this build-up. Ear, nose and throat (ENT) Departments may fit children with artificial valves (grommets) in the membrane to relieve pressure. In addition, the eustachian tube may become blocked (e.g. as a result of swollen adenoids) and air is gradually absorbed into the surrounding mucous membrane creating a partial vacuum.

Within the middle ear are three ossicles (small bones) – the hammer (malleus), anvil (incus) and stirrup (stapes). These form a flexible lever system, which transmit sound waves across the tympanic cavity. The footplate of the stapes makes contact with the small oval window, through which vibrations are transmitted to the inner ear or cochlea. Small muscles attached to the malleus (tensor tympani) and the stapes (stapedius), can stabilize both the tympanic membrane and the oval window, thus giving each sound a definite end point and preventing excessive movement in very loud sounds.

The roof of the middle ear is thin; sometimes infection can be transmitted from the middle ear into the cranial cavity. Large vessels such as the internal jugular vein lie below the middle ear cavity.

The inner ear apparatus contains

(1) the vestibular apparatus, which detect gravity, acceleration and head position; and
(2) the cochlea, which detects sound waves.

Both sets of sense organs are surprisingly similar.

Sound waves cause vibrations of the tympanic membrane (or eardrum); these vibrations pass along the chain of small bones (ossicles) in the middle ear and cause piston-like movements of the footplate of the stirrup which are, in turn, transmitted to the cochlea.

The cochlea: the cochlea or inner ear is a closed, spirally arranged system subdivided by fibrous septa into three channels which contain fluid. The upper and lower channels are the *scala vestibuli* and *scala tympani*, respectively; they contain endolymph, a fluid similar to cerebrospinal fluid. They sandwich between them the *scala media* or cochlear duct which contains the *Organ of Corti*. This highly specialized structure consists of

(1) rows of *hair cells* (whose "hairs" or stereocilia protrude into a gelatinous flap – the *tectorial membrane* – and the surrounding fluid medium of perilymph)
(2) supporting cells such as *phalangeal cells* and *pillar cells*

Sound waves set up vibration of the *basilar membrane* on which the organ of Corti is supported – and thus set up shearing forces between the hairs of the hair cells on the one hand and the tectorial membrane/ surrounding fluid on the other. The *position of maximal amplitude* of the standing wave along the membrane determines the *pitch* of the sound perceived. The *loudness* of the sound alters the area responding at maximal amplitude, and hence the number of cells firing.

Auditory pathways: the auditory pathways are unduly complex. When stimulated by sound, the release of neurotransmitter from the base of the hair cells alters; this stimulates the endings of neurons whose cell bodies lie deep within the petrous temporal bone (collectively the *spiral ganglion*). Inside the brainstem, most fibers decussate (=cross the brainstem) as the *trapezoid body*, and ascend as the *lateral lemniscus*. There are further synaptic stations before the pathway proceeds to the *auditory cortex* in the temporal lobe of the forebrain.

Some fibers in the brainstem do not decussate but, otherwise, seem to follow the same pathways described above. *Damage to one pathway*, therefore, has little effect on hearing; conversely, *damage to one ear* makes it difficult to localize sound, an activity which depends on sound arriving at each ear at slightly different times.

The colliculi are midbrain structures involved in reflex movements in response to sound (inferior colliculus) or vision (superior colliculus), particularly "startle" responses, e.g. where a person turns quickly in response to a noise.

The eighth cranial nerve is the vestibulocochlear nerve. This nerve originates from between the pons and medulla. The vestibulocochlear nerve is a dual nerve with both components being of a special sensory function. The two parts to this nerve are the vestibular component and the cochlear division. The vestibular component deals with information related to balance, with the cochlear nerve relaying information related to hearing. After the nerve components leave the brainstem, it enters the petrous temporal bone via the internal auditory (acoustic) meatus. At this point, it is very closely related to the facial nerve and the labyrinthine vessels. There are several accounts of the communication between the facial and vestibulocochlear nerves at this point (Paturet, 1951; Fisch, 1973; Shoja et al., 2014).

At the lateral edge of the internal auditory meatus, the vestibulocochlear nerve divides into its more anterior cochlear nerve and the posterior vestibular nerve. The cochlear nerve, as its name suggests supplies the cochlea, which transmits information related to hearing. The vestibular nerve however, supplies the utricle and saccule, as well as the ampullary crests of the semicircular ducts. In essence, the vestibular nerve conducts impulses related to balance.

There are two ganglia associated with the vestibulocochlear nerve – the vestibular and spiral ganglia. The vestibular ganglion has three major branches – superior, inferior and posterior. These fibers are distributed as follows.

- *Superior division* – to the macula of the utricle and the ampullary crests of the semicircular ducts (lateral and anterior)
- *Inferior division* – passes to the macula of the saccule
- *Posterior division* – passes to the ampullary crest of the posterior semicircular duct

The ganglion related to the cochlear nerve is called the spiral ganglion. For hearing, the cochlea contains the organ of Corti, also called the spiral organ, hence the origin of the term spiral ganglion. The fibers from the spiral ganglion pass to the spiral lamina and the outer hair cells. The spiral ganglion contains two cell types: (a) large myelinated bipolar cells, for the inner hair cells and (b) small unmyelinated cells for the outer hair cells.

The cells arising from the vestibular ganglion then carry information to the brain above and to the cochlear nerve medially. Those fibers arising from the spiral ganglion, project to the primary auditory cortex in the superior temporal gyrus of the cerebral cortex.

1.5.8.1 Typical Pathologies to Affect the Vestibulocochlear Nerve

A wide variety of pathologies may affect the vestibulocochlear nerve. Deafness can affect the vestibulocochlear nerve by either a conductive or sensorineural deafness. Conductive deafness can present due to earwax simply blocking the transmission of sound impulses down the external auditory meatus, or perhaps due to a foreign body in this canal. Sensorineural deafness can result due to pathology of the cochlea, the cochlear division of this nerve, or perhaps within central pathways.

Vertigo can be expressed by a patient and can be due to a number of causes, originating centrally (like neurological, infections or head trauma) or peripherally (cholesteatoma, labyrinthitis benign postural vertigo or rarely Ménière's disease). Tinnitus, a ringing or buzzing in the ear, can present due to a variety of causes like infections or drug treatments like gentamycin.

An acoustic neuroma, or vestibular schwannoma, is a small growing tumor, generally occurring at the internal auditory meatus. This presents as a sensorineural deafness, but can also present with features affecting the facial nerve, due to its close anatomical relation to this nerve at the site of entry into the petrous temporal bone.

DYSPHASIAS are defects in the understanding and construction of speech. The neural substrate for human speech is located in the *left cerebral hemisphere* in some 97% of the population. Two areas are of importance

- A sensory speech area located near the auditory cortex and covering quite a large area of the temporal and parietal lobes. It has an important core region – Wernicke's area.
- A motor speech area located in the frontal lobe near the face, tongue and larynx regions of the motor cortex – Broca's area.

Damage to Wernicke's or Broca's area (or pathways that link them) leads to dysphasia. Such damage might result from a stroke or a tumor, and might occasionally be widespread enough to affect both areas. If only one area is affected, the results are different depending which of the two is involved.

If Wernicke's area is damaged:
(1) The speech of others is heard (therefore the patient is not deaf) but is not understood. Conversion of sounds to ideas does not occur, and the patient cannot obey instructions
(2) Patients cannot monitor their own speech, which becomes excessive, unintelligible, full of jargon and new made-up words (neologisms).

Patients usually have no insight into their state of deficiency, and may appear so withdrawn from reality that diagnoses of other mental disorders are made.

If Broca's area is damaged:
(1) The patient understands the speech of others and spoken words lead to sensible ideas. However,
(2) Patients cannot find words to express themselves, are non-fluent, hesitant, make grammatical errors, omissions, and may speak like telegrams. They may stammer and, in extreme cases, be rendered mute.

Patients usually have insight into the deficit and become very frustrated.

Because of the way in which damage to the brain occurs, it is unlikely to be restricted to the speech area alone. Damage to the region that includes Wernicke's area may, e.g. also result in difficulties with writing (dyslexia) and numbers, visual impairment (part of the visual pathway, called Meyer's loop, runs through the temporal lobe) and impaired memory.

Damage to the region that includes Broca's area might also result in weakness of the R. face and upper limb (and even sensory loss from these regions) by involving the sensorimotor strip.

Dysphasias may also occur because of damage which disconnects links between Wernicke's and Broca's areas. Do not assume that this is well understood.

DYSARTHRIAS are mechanical defects in speech, arising from difficulties in moving lips, tongue, palate, pharynx, larynx, etc. In other words, clear pronunciation is impaired due to a neuromuscular deficit, but the "sense" of speech is quite normal.

Pseudobulbar palsy involves damage to pathways connecting the cerebral cortex to cranial nerves in the brainstem. These cranial nerves are the motor nerves V (muscles of mastication), VII (facial musculature), IX, X, XI (pharynx and larynx), XII (tongue). Unilateral damage does little to affect speech unless VII is involved, when some slurring occurs. Bilateral damage does disturb speech; it is most likely to result from some global effect on nerves, such as multiple sclerosis, motor neuron disease, etc.

Bulbar palsy usually involves damage to the brainstem itself (and hence to the cranial nerves referred to above which are located within the brainstem). However, it is a loose term sometimes applied to damage to the nerve axons or neuromuscular junctions, or to the muscles they supply. In any event, speech defects would be the least of worries; there would be difficulty in swallowing (leading to aspiration of food and fluids and pneumonia) and, probably, difficulties in respiration. Bulbar palsy is life threatening.

Dysarthrias also occur because of damage to other brain regions which control movement, such as the basal ganglia or the cerebellum. Damage to the basal ganglia may result in quiet and unemotional (deadpan) speech, while cerebellar damage may result in "drunken" speech.

DYSPHONIA is quiet or hoarse speech resulting from abnormalities in the passage of air across the vocal folds. It might result from

(1) infections or pathological changes such as laryngitis, the development of nodules on the vocal folds, or carcinoma of the larynx;
(2) damage to the innervation of the laryngeal muscles; and
(3) weakness of the chest muscles.

1.5.8.2 Clinical Testing of the Vestibulocochlear Nerve

Clinical examination of this cranial nerve should incorporate a basic test of hearing perhaps at the bedside, Weber's and Rinne's test (though these can be unpopular means of clinical examination) and otoscopy. More advanced and specific testing may be needed and can include automated otoacoustic emission tests, automated auditory brainstem response testing, pure tone audiometry assessment or bone conduction testing. These advanced tests will assess hearing, but it may also be appropriate to test the vestibular system by rotation testing, electronystagmography, computerised dynamic posturography, vestibular evoked myogenic potential or CT/MRI scanning of the brain. Clinical examination of this nerve, and related further explanations of the anatomy and pathologies of the vestibulocochlear nerve, can be found in Clinical Anatomy of the Cranial Nerves (Rea, 2014), the companion text to this book. A brief discussion is contained in Chapter 10.

1.5.9 Glossopharyngeal Nerve

The ninth cranial nerve is the glossopharyngeal nerve, and contains many different types of fibers within it. The glossopharyngeal nerve is comprised of branchial motor, visceral motor, and special and general sensory fibers. It contains sensory fibers from the pharynx, tongue (posterior one-third) and the tonsils. Secretomotor fibers destined for the parotid gland as well as motor fibers for the stylopharyngeus are transmitted with the glossopharyngeal nerve. Finally, it contains taste fibers from the posterior one-third of the tongue.

Originating at the medulla oblongata, it then passes intimate to the olive and inferior cerebellar peduncle. It passes through the jugular foramen, where its superior and inferior ganglia are also found. At the point it passes through the jugular foramen, it is closely related to the internal jugular vein and internal carotid artery.

The glossopharyngeal nerve has five major components – special and general visceral efferent, general and special visceral afferent, and general somatic afferent.

- *Special visceral efferent*: this component of the glossopharyngeal nerve transmits information to the stylopharyngeus muscle, a muscle originating from the third pharyngeal arch. This muscle is responsible for raising the larynx and pharynx, and functions during swallowing.

- *General visceral efferent*: this component of the glossopharyngeal nerve transmits the parasympathetic information to the parotid salivary gland. Although the facial nerve goes through the parotid gland, it is not that nerve that innervates it, but the glossopharyngeal nerve.
- *General visceral afferent*: this component of the glossopharyngeal nerve transmits sensory information from the carotid body and sinus, middle ear, posterior one-third of the tongue and vallate papillae, mucous membranes of the soft palate and palatine tonsils and the oropharynx.
- *Special visceral afferent*: this component of the glossopharyngeal nerve transmits information related to taste from the posterior one-third of the tongue. Remember that taste fibers from the front two-thirds of the tongue are carried in the facial nerve.
- *General somatic afferent*: this component of the glossopharyngeal nerve transmits general sensory information from inside of the tympanic membrane, skin of the external ear, upper portion of the pharynx and general sensation from the posterior one-third of the tongue.

The glossopharyngeal nerve has two ganglia associated with it – a superior and inferior ganglion. The superior ganglion transmits visceral sensory fibers from the carotid sinus and body, pharynx and parotid gland and the middle ear. The inferior ganglion, the larger of the two, conveys two types of information in it – general and special sensory. This information arises from the posterior one-third of the tongue.

1.5.9.1 Typical Pathologies to Affect the Glossopharyngeal Nerve

An isolated glossopharyngeal nerve palsy is very rare. If one glossopharyngeal nerve does suffer from palsy, it tends not to result in any major deficit for the patient due to corticobulbar input bilaterally. However, pseudobulbar palsy can result which is a condition that causes degeneration of cranial nerve nuclei bilaterally, and the corticobulbar tracts (the pathway which connects the brainstem and the cerebral cortex, discussed in more detail in Chapter 9). It would present as difficulty in swallowing (dysphagia) and problematic production of speech, i.e. affecting the motor aspect of speech.

1.5.9.2 Clinical Testing of the Glossopharyngeal Nerve

Testing of the glossopharyngeal nerve can be undertaken at the bedside quite simply by listening to the patient's voice, and asking them to say "ahh" without tongue protrusion. If there is pathology of the glosso-pharyngeal nerve, the uvula would *be pulled to the normal (unaffected) side*. Further details on the pathologies to affect the glossopharyngeal nerve and the clinical examination of it are covered in Clinical Anatomy of the Cranial Nerves (Rea, 2014), the companion text to this book. A brief discussion is contained in Chapter 10.

1.5.10 Vagus Nerve

The tenth cranial nerve is the vagus nerve. It is a very complex nerve and has five different types of fibers within it. These are summarized below.

(1) Branchial motor
Supplying all muscles of the pharynx and larynx, except stylopharyngeus (which is supplied by the glossopharyngeal nerve) and tensor veli palatini (which is supplied by the medial pterygoid nerve from the mandibular division of the trigeminal nerve).

(2) Visceral sensory
This component of the vagus nerve is responsible for transmitting information from a wide variety of anatomical sites including the heart and lungs, pharynx and larynx and upper part of the gastrointestinal tract.

(3) Visceral motor
The visceral motor component carries parasympathetic fibers from the smooth muscle of the upper respiratory tract, heart and gastrointestinal tract.

(4) Special sensory
The special sensation conveyed by the vagus nerve is for taste from the palate and epiglottis.

(5) General sensory
The general sensory component of the vagus nerve is concerned with information from parts of the ear and the dura within the posterior cranial fossa.

The vagus nerve arises from the level of the medulla. It then passes through the jugular foramen between the glossopharyngeal and spinal accessory nerves. There is a variation between the left and right side of the body of the course of each vagus nerve.

On the left side, the vagus nerve passes down to the thorax between the subclavian and common carotid arteries, behind the brachiocephalic vein. After providing branches to the heart, lungs and esophagus, the left vagus nerve then passes to the arch of the aorta. The recurrent laryngeal nerve is then giving off, which passes under the arch of the aorta then ascends to supply the lower aspect of the larynx. During its course, it also provides further supply to the heart, aorta, esophagus and trachea.

On the right side, the vagus nerve passes posterior to the superior vena cava and anterior to the right subclavian artery. At the point of the subclavian artery, it gives off the recurrent laryngeal nerve supplying the upper portion of the larynx.

1.5.10.1 Typical Pathologies to Affect the Vagus Nerve
As with the glossopharyngeal nerve, an isolated vagus nerve palsy is extremely rare. However, if the vagus nerve is affected by trauma or neurological disease, the effects are wide and varied, due to the large numbers of structures it supplies. Detailed below are a few pathologies that affect the vagus nerve, but further details are discussed in the companion text to this book in Clinical Anatomy of the Cranial Nerves (Rea, 2014).

(1) Pseudobulbar palsy
    A pseudobulbar palsy is caused by a wide variety of conditions but typically results from bilateral degeneration involving cranial nerve nuclei and the corticobulbar tract (pathway connecting the brainstem with the cerebral cortex). It results in the patient having dysphagia and difficulty in the motor aspect of speech production.
(2) Bilateral vagus nerve nucleus pathologies
    A bilateral pathology affecting the vagus nerve will result in paralysis of the pharynx and larynx.
(3) Injury to the recurrent laryngeal nerve
    The left recurrent laryngeal nerve runs a slightly longer course and tends to be affected more than the right for pathologies. An aneurysm of the aorta can result in compression of the left recurrent laryngeal nerve. In addition, any neck operation will place the recurrent laryngeal nerve at risk, especially if the operative field is close to the trachea-esophageal groove, heart, lungs or esophagus. In addition, thyroidectomy can put the

recurrent laryngeal nerve at risk from damage, but this occurs in approximately 1% of individuals. If one recurrent laryngeal nerve is damaged, it will result in dysphonia (difficulty with speech) and hoarseness. If there is bilateral recurrent laryngeal nerve damage, it can present as a surgical emergency with inspiratory stridor, aphonia and laryngeal obstruction. It may need to be treated by a tracheostomy in the first instance.

(4) Injury to the superior laryngeal nerve

Injury to the superior laryngeal nerve can occur as a complication of a thyroidectomy. It will result in paralysis of the cricothyroid muscle and anesthesia of the region above the level of the vocal folds. It tends to be, however, the external laryngeal branch that is affected. Therefore, it would affect only the cricothyroid muscle. Some patient's may not have any significant consequences of this, while others may have difficulty in changing the pitch of their voice, ore reduced stamina in their voice.

1.5.10.2 Clinical Testing of the Vagus Nerve

As with testing of the glossopharyngeal nerve, testing of the vagus nerve is done simply by listening to the patient's voice, and asking them to say "ahh". Normally, the soft palate would rise equally in the midline. However, a vagus nerve palsy (or indeed a glossopharyngeal nerve palsy) would result in would result in the uvula *being pulled to the normal (unaffected) side.* As with testing of the glossopharyngeal nerve, the integrity of the vagus nerve may be assessed by testing the gag reflex. Obviously, the conscious patient must be told of what is happening and being assessed. A brief discussion of the clinical assessment of the vagus nerve is presented in Chapter 10.

1.5.11 Spinal Accessory Nerve

The eleventh cranial nerve is the spinal accessory nerve. The spinal accessory nerve is a purely somatic motor nerve. It innervates only two muscles – the trapezius and the sternocleidomastoid muscles. In addition, the spinal part, although generally referred to as a somatic motor nerve supplying the trapezius and sternocleidomastoid muscles, also contains sensory fibers similar from stretch receptors (Echlin and Propper, 1938), non-proprioceptive sensory and nociceptive fibers (Bremner-Smith et al., 1999). However, its cranial portion supplies the muscles of the soft palate.

The spinal accessory nerve is comprised of two different parts – the spinal and cranial portions. The cranial part is not present in every person (Tubbs et al., 2014), but generally arises from the vagus nerve. This cranial part then will join with the spinal part of the accessory nerve, which arises from the first five or six cervical spinal nerves. The spinal portion then ascends into the foramen magnum passing laterally to join with the cranial root.

Both parts of the accessory nerve then pass into the jugular foramen, alongside the vagus and glossopharyngeal nerves. The cranial root then passes to the vagus nerve's superior ganglion. The spinal portion then passes inferiorly to supply its two related muscles – the sternocleidomastoid and trapezius muscles.

1.5.11.1 Typical Pathologies to Affect the Spinal Accessory Nerve

Lesions of the accessory nerve are divided into where anatomically the pathology arises from. It is divided into supranuclear lesions, compression at the jugular foramen and lesions within the posterior triangle.

A supranuclear lesion will affect both sternocleidomastoid and trapezius muscles bilaterally. If occlusion happens to the nearby vasculature, e.g. the posterior inferior cerebellar artery, it can affect several of the cranial nerves. Specifically, it will affect the glossopharyngeal, vagus and the accessory nerves. Many areas of the head and neck will be affected, as it will then result in the gag reflex and speech being affected, as well as balance and sensation on the face being affected. This condition is referred to as Wallenberg's syndrome.

Jugular foramen pathology will result in the spinal accessory nerve being affected, If the lesion is large enough, e.g. neoplasia, or vascular anomaly, it can also compress the other nearby cranial nerves, i.e. glossopharyngeal and vagus nerves. This would result in the patient having weakness or paralysis of the trapezius and sternocleidomastoid muscles, paralysis of the vocal cords, anesthesia of the larynx and pharynx, as well as dysphagia. This condition is called jugular foramen syndrome, or Vernet's syndrome.

Within the posterior triangle, the spinal accessory nerve may be damaged due to surgical procedures like a radical neck dissection for complete clearance of cancer in that region, or perhaps even biopsy of tissue in that territory for assessment of lymph nodes.

1.5.11.2 Clinical Testing of the Spinal Accessory Nerve

From the clinical perspective, the assessment of the spinal accessory nerve is undertaken by assessment of the function of sternocleidomastoid and trapezius.

You have to keep in mind that the sternocleidomastoid serves two purposes, dependent on if it is acting on its own, or both left and right muscles contract. Each sternocleidomastoid muscle rotates the head to the opposite side and slightly raises the head. Both the left and right sternocleidomastoid muscles contracting results in flexion of the neck. Testing for the integrity of the muscle will involve rotating the head to the opposite side of testing and having the patient "push back" their neck into the examiners hand. This not only assesses the function of the muscle, but also its strength.

Trapezius also has a number of functions, dependent on which fibers are working. The upper fibers will raise the scapula, the middle fibers will pull it medially and the lowermost fibers will pull down the scapula. Clinical assessment of the trapezius is undertaken by asking the patient to raise their shoulders and resist pressing down on the shoulders, providing the patient does not suffer any arthritic changes in that territory, as we should not cause harm or hurt to the patient.

Further testing of the spinal accessory nerve will be directed by the clinical presentation and can include, but not limited to, electromyographic studies and/or neck/cranial CT/MRI scanning. Further details on related pathologies and clinical examination of the spinal accessory nerve can be found in Clinical Anatomy of the Cranial Nerves (Rea, 2014), the companion text to this book. A brief discussion of clinical examination is contained in Chapter 10.

1.5.12 Hypoglossal Nerve

The twelfth cranial nerve is the hypoglossal nerve and is purely a somatic motor nerve. It supplies all but one (palatopharyngeus, supplied by the glossopharyngeal nerve) of the intrinsic and extrinsic muscles of the tongue. There are some general sensory fibers which hook onto the hypoglossal nerve, and they originate from the second cervical nerve.

The hypoglossal nerve arises from the hypoglossal nucleus located the full length of the medulla. The two major bundles of the hypoglossal nerve then arise from between the pyramid and olive and pass into the hypoglossal canal. On exiting the hypoglossal canal, the hypoglossal nerve then passes behind the internal carotid artery. It passes around the occipital artery and passes over the external and internal carotid arteries as well as the lingual artery. It then passes to the floor of the mouth, intimately associated with the hyoglossus and below the lingual nerve and duct of the submandibular gland.

1.5.12.1 Typical Pathologies to Affect the Hypoglossal Nerve

Like the spinal accessory nerve, pathologies affecting the hypoglossal nerve can be subdivided into those that are supranuclear, pathologies of the hypoglossal nucleus itself, and those which affect the nerve in its course to supply the muscles of the tongue.

Supranuclear pathology of the hypoglossal nerve is very variable in their presentation, but will result in a mild weakness, which may come and go of the tongue musculature of the affected side.

Pathology of the hypoglossal nucleus itself generally would affect the midline structures. As the hypoglossal nerve and its nucleus arise so close to the midline, pathology of one side can affect the opposite side too. A variety of causes can affect the hypoglossal nucleus and can include tumors of the spinal cord, or nearby structures compressing it, or perhaps infarction of nearby vasculature.

Pathology of the hypoglossal nerve can result anywhere along its course to the muscles of the tongue. Typically, this would involve a tumor along its course either intracranially or extracranially. The hypoglossal nerve can also be affected by neurological pathology, trauma or infections like meningitis.

1.5.12.2 Clinical Testing of The Hypoglossal Nerve

Clinical examination of the hypoglossal nerve is covered in more detail in the companion book to this one, Clinical Anatomy of the Cranial Nerves (Rea, 2014). However, briefly, the hypoglossal nerve, and the structures it supplies, would result in the tongue pointing *toward the side of the lesion*. This is because the tongue would be "pushed over" toward the pathological side due to the unopposed power of the tongue on the unaffected side.

1.5.13 Summary of Cranial Nerves

The following table (Table 1.4) summarizes all the key features in relation to all of the 12 pairs of the cranial nerves. It details the main com-

Table 1.4. Summary Table of the Cranial Nerves Detailing the Salient Features of Each Nerve

Nerve	Components	Functions	Point of Entry/ Exit from Brain	Exits/Enters Cranial Cavity	Nuclei	Ganglion	Important Branches
Olfactory (I)	Special sensory	Smell	Forebrain	Cribriform plate of ethmoid bone	No specific nucleus. Olfactory epithelium contain the cell bodies	None	Olfactory epithelium (central processes)
Optic (II)	Special sensory	Vision	Midbrain	Optic canal	Lateral geniculate nucleus	Retinal ganglion cells	Optic nerve; optic tract
Oculomotor (III)	Somatic motor Visceral motor	Extraocular muscles Sphincter muscle and ciliary muscle	Midbrain	Superior orbital fissure	Oculomotor nucleus; Edinger–Westphal nucleus	Ciliary ganglion	Motor branches to extraocular muscles; Parasympathetic division
Trochlear (IV)	Somatic motor	Innervates the superior oblique muscle	Midbrain	Superior orbital fissure	Nucleus of the trochlear nerve	None	None. Only supplies the superior oblique muscle
Trigeminal (V)	General sensory; Branchial motor	Sensation from face; paranasal sinuses; nose and teeth Muscles of mastication	Pons	Superior orbital fissure (Va), foramen rotundum (Vb) or foramen ovale (Vc)	Spinal trigeminal nucleus; Pontine trigeminal nucleus; Mesencephalic trigeminal nucleus; Trigeminal motor nucleus	Trigeminal ganglion; submandibular ganglion	Ophthalmic nerve; Maxillary nerve; Mandibular nerve
Abducent (VI)	Somatic motor	Innervates the lateral rectus muscle	Pontomedullary junction	Superior orbital fissure	Abducent nerve nucleus	None	None. Only supplies the lateral rectus muscle

Facial (VII)	Branchial motor Visceral motor Special sensory General sensory	Muscles of facial expression, stylohyoid, stapedius, posterior belly of digastric Parasympathetic innervation of the submandibular and sublingual salivary glands, lacrimal gland and the nasal and palatal glands Anterior two-thirds of the tongue (and palate) Concha of the auricle	Pontomedullary junction	Stylomastoid foramen	Facial motor nucleus; lacrimal nucleus; superior salivatory nucleus; Gustatory nucleus; Spinal trigeminal nucleus	Geniculate ganglion; pterygopalatine ganglion; submandibular ganglion	**Intratemporal** Greater petrosal nerve; nerve to stapedius; chorda tympani **Extratemporal** Temporal; zygomatic; buccal; marginal mandibular; cervical; posterior auricular; posterior belly of digastric branch; stylohyoid branch
Vestibulocochlear (VIII)	Special sensory	Balance for the vestibular component; Hearing for the spiral (cochlear) component	Pontomedullary junction	Internal auditory meatus	Vestibular nucleus; ventral cochlear nucleus; dorsal cochlear nucleus; superior olivatory nucleus	Vestibular ganglion; Spiral ganglion	Vestibular nerve; cochlear nerve
Glosso-pharyngeal (IX)	Branchial motor Visceral motor Special sensory General sensory Visceral sensory	Stylopharyngeus Parotid gland for parasympathetic innervation Taste from the posterior one-third of the tongue External ear Pharynx; parotid gland; middle ear; carotid sinus and body	Medulla oblongata	Jugular foramen	Nucleus ambiguus; Solitary nucleus; spinal trigeminal nucleus; Inferior salivatory nucleus	Inferior ganglion; Otic ganglion; Superior ganglion; inferior ganglion	Muscular; tympanic; pharyngeal; tonsillar; carotid sinus branch

(Continued)

Table 1.4. Summary Table of the Cranial Nerves Detailing the Salient Features of Each Nerve *(cont.)*

Nerve	Components	Functions	Point of Entry/ Exit from Brain	Exits/Enters Cranial Cavity	Nuclei	Ganglion	Important Branches
Vagus (X)	Branchial motor Visceral motor Special sensory General sensory Visceral sensory	Pharyngeal constrictors; laryngeal muscles (intrinsic); palatal muscles; upper two-thirds of esophagus heart; trachea and bronchi; gastrointestinal tract Taste from the palate and the epiglottis Auricle; external auditory meatus; posterior cranial fossa dura mater Gastrointestinal tract (to last one-third of the transverse colon); pharynx and larynx; trachea and bronchi; heart	Medulla oblongata	Jugular foramen	Dorsal nucleus of the vagus nerve; nucleus ambiguus; soliatry nucleus; spinal trigeminal nucleus	Superior ganglion; inferior ganglion	Meningeal branch; auricular branch; pharyngeal branches; superior laryngeal nerve; recurrent laryngeal nerve; carotid branches; cardiac branches; esophageal branches; pulmonary branches; gastric branches; celiac branches; renal branches
Spinal Accessory (XI)	Somatic motor	Innervates the sternocleidomastoid and trapezius muscles	Medulla oblongata (and spinal cord)	Jugular foramen	Nucleus ambiguus; Spinal accessory nucleus	None	Cranial branch; spinal branch
Hypoglossal (XII)	Somatic motor	Extrinsic and intrinsic muscles of the tongue. Palatoglossus is not supplied by the hypoglossal nerve. It is supplied by the glossopharyngeal nerve	Medulla oblongata	Hypoglossal canal	Hypoglossal nucleus	None. It may however receive general sensory fibers from the ganglion of C2	Meningeal branches; Thyrohyoid branches; Muscular branches;

ponents within each nerve, its function(s), point of entry or exit from the brain, the aperture it enters or leaves the cranial cavity, its related nuclei (if any are associated with it), ganglion or ganglia (again, if associated with the nerve) and its important branches, especially from an anatomical and functional perspective.

REFERENCES

Afifi, A., Bergman, R., 2005. Functional Neuroanatomy, second ed. McGraw-Hill, USA.

Benzel, E.C., 2012. The Cervical Spine, fifth ed. Lippincott Williams and Wilkins, Philadelphia, USA.

Bester, H., Chapman, V., Besson, J.M., Bernard, J.F., 2000. Physiological properties of the lamina I spinoparabrachial neurons in the rat. J. Neurophysiol. 83, 2239–2259.

Bremner-Smith, A.T., Unwin, A.J., Williams, W.W., 1999. Sensory pathways in the spinal accessory nerve. J. Bone Joint Surg. (Br.) 81, 226–228.

Cabot, J.B., Alessi, V., Carroll, J., Ligorio, M., 1994. Spinal cord lamina V and lamina VII interneuronal projections to sympathetic preganglionic neurons. J. Comp. Neurol. 347, 515–530.

Christensen, B.N., Perl, E.R., 1970. Spinal neurons specifically excited by noxious or thermal stimuli:marginal zone of the dorsal horn. J. Neurophysiol. 33, 293–307.

Craig, A.D., Kniffki, K.D., 1985. Spinothalamic lumbosacral lamina I cells responsive to skin and muscle stimulation in the cat. J. Physiol. 365, 197–221.

Echlin, F., Propper, N., 1938. Sensory fibres in the spinal accessory nerve. J. Physiol. 92, 160–166.

Fajardo, C., Escobar, M.I., Buriticá, E., Arteaga, A., Umbarila, J., Casanova, M.F., Pimienta, H., 2008. Von Economo neurons are present in the dorsolateral (dysgranular) prefrontal cortex of humans. Neurosci. Lett. 435 (3), 215–218.

Fisch, U.P., 1973. Excision of the Scarpa's ganglion. Arch. Otolaryngol. 97, 147–149.

Honda, C.A., Lee, C.L., 1985. Immunohistochemistry of synaptic input and functional characterisation of neurons near the spinal central canal. Brain Res. 343,120–128.

Honda, C.N., 1985. Visceral and somatic efferent convergence onto neurons near the central canal in the sacral spinal cord of the cat. J. Neurophysiol. 53, 1059–1078.

Honda, C.N., Perl, E.R., 1985. Functional and morphological features of neurons in the midline region of the caudal spinal cord of the cat. Brain Res. 340, 285–295.

International Standards for Neurological Classification of Spinal Cord Injury (ISNCSCI). American Spinal Injury Association. http://www.asia-spinalinjury.org/elearning/ASIA_ISCOS_high.pdf (accessed 28.05.2014).

Ju, G., Hökfelt, T., Brodin, E., Fahrenkrug, J., Fischer, J.A., Frey, P., Elde, R.P., Brown, J.C., 1987a. Primary sensory neurons of the rat showing calcitonin gene-related peptide immunoreactivity and their relation to substance P, somatostatin, galanin, vasoactive intestinal polypeptide and cholecystokinin immunoreactive ganglion cells. Cell Tissue Res. 247, 417–431.

Ju, G., melander, T., Ceccatelli, S., Hökfelt, T., Frey, P., 1987b. Immunohistochemical evidence for a spinothalamic pathway co-containing cholecystokinin and galanin like immunoreactivities in the rat. Neuroscience 20, 439–456.

Kiernan, J.A., Rajakumar, N., 2014. Barr's The Human Nervous System. An Anatomical View-point, tenth ed. Lippincott Williams and Wilkins, Philadelphia, USA, ISBN 145117327X.

Menétrey, D., Giesler, Jr., G.J., Besson, J.M., 1977. An analysis of response properties of spinal dorsal horn neurons to non-noxious and noxious stimuli in the rat. Exp. Brain Res. 27, 15–33.

Nahin, R.L., Madsen, A.M., Giesler, G.J., 1983. Anatomical and physiological studies of the grey matter surrounding the spinal cord central canal. J. Comp. Neurol. 220, 321–335.

NHS, Treating Trigeminal Neuralgia. http://www.nhs.uk/Conditions/Trigeminal-neuralgia/Pages/Treatment.aspx (accessed 11.05.2014).

Nicholas, A.P., Zhang, X., Hökfelt, T., 1999. An immunohistochemical investigation of the opioid cell column in lamina X of the male rat lumbosacral spinal cord. Neurosci. Lett. 270, 9–12.

Paturet, G., 1951. Traite D'anatomie Humainevol. 1Masson et Cie, Paris.

Rea, P., 2014. Clinical Anatomy of the Cranial Nerves, first ed. Academic Press, Elsevier, San Diego, USA.

Renshaw, B., 1946. Central effects of centripetal impulses in axons of spinal ventral roots. J. Neurophysiol. 9, 191–204.

Rexed, B., 1952. The cytoarchitectonic organisation of the spinal cord in the cat. J. Comp. Neurol. 96, 414–495.

Shoja, M.M., Oyesiku, N.M., Griessenauer, C.J., Radcliff, V., Loukas, M., Chern, J.J., Benninger, B., Rozelle, C.J., Shokouhi, G., Tubbs, R.S., 2014. Anastomoses between lower cranial and upper cervical nerves. Clin. Anat. 27, 118–130.

Siegel, G.J., Agranoff, B.W., Albers, R.W., Fisher, S.K., Uhler, M.D. (Eds.), 1999. Basic Neurochemistry: Molecular, Cellular and Medical Aspects. Lippincott-Raven, Philadelphia, USA, ISBN 039751820X.

Siegel, A., Sapru, H.N., 2006. Essential Neuroscience. Lippincott Williams and Wilkins, Baltimore.

Snyder, R.L., Faull, R.L., Mehler, W.R., 1978. A comparative study of the neurons of origin of the spinocerebellar afferents in the rat, cat and squirrel monkey based on the retrograde transport of horseradish peroxidase. J. Comp. Neurol. 15, 833–852.

Tubbs, R.S., Benninger, B., Loukas, M., Gadol, A.A.C., 2014. Cranial roots of the accessory nerve exist in the majority of adult humans. Clin. Anat. 27, 102–107.

Todd, A.J., 1996. GABA and glycine in synaptic glomeruli of the rat spinal dorsal horn. Eur. J. Neurosci. 8, 2492–2498.

Todd, A.J., Lewis, S.G., 1986. The morphology of Golgi-stained neurons in lamina II of the rat spinal cord. J. Anat. 149, 113–119.

Walker, H.K., Hall, W.D., Hurst, J.W., 1990. Clinical Methods. The History, Physical, and Laboratory Examinations, third ed. Butterworths, Boston, ISBN-10: 0-409-90077-X. Found at: http://www.ncbi.nlm.nih.gov/books/NBK201/ (accessed 11.05.2014).

Wang, Y., Gupta, A., Toledo-Rodriguez, M., Wu, C.Z., Markram, H., 2002. Anatomical, physiological, molecular and circuit properties of nest basket cells in the developing somatosensory cortex. Cereb Cortex. 12 (4), 395–410.

Willis, W.D., Coggeshall, R.E., 1991. Sensory Mechanisms of the Spinal Cord, second ed. Plenum Press, New York.

CHAPTER 2

Essential Anatomy and Function of the Brain

2.1 INTRODUCTION

The skull is a logical consequence of the process of cephalization – an evolutionary trend in which multicellular organisms have developed an elongated, cylindrical body with a leading (front) and trailing (back) end. There are clear advantages in grouping sense organs at the front, and also in making the front end the site of ingestion (food, air and water); to cope with these sense organs, the nervous system enlarges at the front (i.e. a brain is developed) and there is a simultaneous requirement to protect these valuable acquisitions – hence the skull.

The skull is composed of a number of individual bones, but is best appreciated as a totality. It can conveniently be divided into three parts.

(1) The *cranium* which is in turn divided into
    (a) the *neurocranium*, essentially a rounded container for the brain and special senses, and
    (b) the *viscerocranium*, the irregularly shaped part at the front which provides a skeletal framework for the nasal and oral cavities.
(2) the *mandible*

The neurocranium consists of the rounded vault of the skull and a rather irregular floor, the base of the skull. The bones of the vault are large and easily identified. It is comprised of the frontal bone, two parietal bones and the occipital bone and the two temporal bones at the side. The bones meet (*articulate*) at wavy lines called sutures. These are the coronal, sagittal and lambdoid sutures. These are immobile, fibrous joints. The newborn baby has gaps between many of the bones called fontanelles (the "soft spots") where the bony plates have yet to grow and meet. The bones of the vault actually develop within a membranous sheet and it is this sheet which fills the interval. The anterior fontanelle is a particularly large, diamond-shaped space which does not fully

Essential Clinical Anatomy of the Nervous System. http://dx.doi.org/10.1016/B978-0-12-802030-2.00002-9

close until *two years after birth*. If the baby is dehydrated, the surface of the head appears to sink in at this point.

The frontal, parietal and much of the occipital bones are covered in life by the scalp, which can be moved backwards and forwards by the occipito-frontalis muscle. The scalp has a rich blood and nerve supply and is such a dense, taut structure that, when wounding occurs, the cut ends of blood vessels tend to pull apart. Thus, scalp wounds bleed profusely and injury often appears worse than it really is.

The sides of the neurocranium are chiefly composed of the temporal bone. This is an extremely complex bone and it is chiefly the flattened squamous part which can be seen from the outside of the skull (a projection from this makes a contribution to the zygomatic arch (cheek bone). In life the temporal bone gives origin to (and is covered by) the temporalis muscle.

The occipital bone can be followed downwards to form part of the base of the skull. This ventral part of the bone has two important features (i) the foramen magnum ("big hole") through which the spinal cord and brain are continuous and (ii) the rounded occipital condyles, by which the skull articulates with the first cervical vertebra – the atlas.

The floor of the interior of the skull of the neurocranium has three levels or "steps", the highest at the front and the lowest at the back. These are the floors of the three cranial fossae (anterior, middle and posterior). The anterior cranial fossa houses the frontal lobes of the brain and is chiefly formed by the frontal bone, especially the orbital plate (which, as its name implies, forms the roof of the orbit). Behind this is the lesser wing of the sphenoid and, in the midline, the cribriform (sieve-like) plate of the ethmoid bone; this is perforated by many holes which transmit olfactory nerves from the nasal cavity which lies beneath it.

The floor of the middle cranial fossa is chiefly formed by the sphenoid bone (which lies in the central part of the fossa) and the temporal bone (which lies at the side). For the sphenoid bone, it comprises the greater wing and the body; the latter has a concave region where the pituitary gland is situated called – rather fancifully – the sella turcica ("turkish saddle"). There are a number of foramina in (or at the edges of) the sphenoid bone which transmit cranial nerves. On each are:

(1) the optic canal (optic nerve and ophthalmic artery);
(2) the superior orbital fissure (ophthalmic veins, nerves to extraocular muscles, ophthalmic division of trigeminal);
(3) the foramen rotundum (maxillary division of trigeminal); and
(4) the foramen ovale (mandibular division of trigeminal).

The floor of the posterior cranial fossa is chiefly formed by the occipital bone. Also present are:

(1) the jugular foramen (through which the glossopharyngeal (IX), vagus (X) and accessory (XI) nerves leave the skull, together with the sigmoid venous sinus which forms the internal jugular vein) and
(2) the hypoglossal canal (through which the hypoglossal nerve leaves the skull).

The brain is at the center of the nervous system for all invertebrate animals and most invertebrate ones too. The anatomy of the human body has been studied since the Egyptian times (1700BC). The first major piece of scientific written work was the Edwin Smith Papyrus and was a surgical text of 48 cases. It included details on injuries, clinical examination, diagnosis, treatment and also prognosis. This was such a significant text because it did not depend on supernatural belief, but rather allowed for a comparative anatomical examination of these features (NIH, 2014).

Since that time, there have been many major contributors to anatomical knowledge and understanding throughout the passage of time. Hippocrates of Cos, during the fourth century BC, provided more evidence for our understanding of the musculoskeletal system and endeavored to enhance our appreciation of the works of the kidney. Then into the fourth century BC, Aristotle changed our viewpoints on many aspects of science due to the philosophical presentation of his thoughts and understanding.

Moving into the third century BC, Herophilos was then viewed as the first anatomist and performed his research on cadavers. He was the first to develop our understanding of the brain as it is today anatomically. He differentiated between the cerebrum and cerebellum, and hypothesized that each part of the brain performed different roles. He also was the first individual to try to determine the function of the optic and oculomotor nerves, and related them to a role in vision, something true to this day.

Nowadays, and with the advance of the field of neuroscience, our understanding of the nervous system has been enhanced due to a variety of techniques employed in the laboratory, e.g. confocal microscopy, stereotaxic injections, tract-tracing studies and immunocytochemistry.

The human brain has been defined as an incredibly complex organ, and we are still gaining new evidence in relation to its functions, and what happens pathologically. The "average" human adult brain comprises approximately 2% of body weight, and its weight can range from 1.2 to 1.4 kg. The brain is an incredibly demanding organ, and is essential for life and function of our body. As such, it consumes an enormous amount of the circulating blood volume. Approximately one-sixth of all cardiac output passes through the brain at any one time, and it uses about one-fifth of all the oxygen in the body when we are at rest.

The brain is our most complex organ and controls and regulates our body, responds to stress and threat, and controls higher cognitive functions. It maintains body temperature, allows us to interpret the special senses, and to socially interact. It ensures the body works optimally in the environment we are in both protecting and nurturing the human body.

2.1.1 Neurons

The brain is composed of two main cell types – neurons and supporting glial cells. Neurons are cells which are *electrically excitable* and transmit information from one neuron to another by chemical and electrical signals. There are three very broad classifications of neurons – *sensory* (which process information on light, touch and sound to name some sensory modalities), *motor* (supplying muscles) and *interneurons* (which interconnect neurons via a network).

Typically a neuron comprises a few basic features, but a few specialisations exist in some areas of the nervous system. In general, a neuron has a cell body. Here, the nucleus – the powerhouse – of the neuron lies with its cytoplasm. At this point, numerous fine fibers enter called dendrites. These processes receive information from adjacent neurons keeping it up-to-date with the surrounding environment. This way the amount of information that a single neuron receives is significantly increased. From a neuron, there is a long single process of variable length called an axon. This conducts information away from the neuron. Some neurons, however, have no axons and the dendrites will conduct information to and from the neuron. In addition to this, a lipoprotein layer called the *myelin sheath* can surround the axon of a principal cell. This is not a continuous layer along the full length of the axon. Rather, there are interruptions called *nodes of Ranvier*. It is at this point where the voltage gated channels occur, and it is at that point where conduction occurs. Therefore, the purpose of the myelin sheath is to enable almost immediate conduction between one node of Ranvier and the next ensuring quick communication between neurons.

In relation to the size of neurons, this varies considerably. The smallest of neurons can be as small as 5 μm, and the largest, e.g. motor neurons can be as big as 135 μm. In addition, axonal length can vary considerably too. The shortest of these can be 100 μm, whereas a motor axon supplying the lower limb, e.g. the toes, can be as long as 1 m.

In the peripheral nervous system, neurons are found in *ganglia*, or in *laminae* (layers) or *nuclei* in the central nervous system.

Neurons communicate with each other at a point called a synapse. Most of these junctional points are chemical synapses where there is the release of a neurotransmitter which diffuses across the space between the two neurons. The other type of synapse is called an electrical synapse. This form is generally more common in the invertebrates, where there is close apposition of one cell membrane and the next, i.e. at the pre and postsynaptic membranes. Linking these two cells is a collection of tubules called *connexons*. The transmission of impulses occurs in both directions and very quickly. This is because there is no delay in the neurotransmitter having to be activated and released across the synapse. Instead, the flow of communication depends on the membrane potentials of the adjacent cells.

The big question that people want an answer is how many cells, or neurons, are present in the brain? Within the CNS, there are many, many millions of neurons. The widely accepted number for the quantity of neurons in the human brain has been approximately 100 billion neurons, with many more supporting cells, or neuroglia. However, recently this idea that there are many more times (up to 10 times) neuroglia compared to neurons has been challenged and may even be approximately the same (Azevedo et al., 2009). It is, however, extremely difficult to put an exact number on this and only an approximation has been possible.

2.1.2 Glial Cells

In addition to neurons, there are supportive cells within the central nervous system too. Their main purpose is in providing nutrient support, maintenance of homeostasis and the production of the myelin sheath. There are two broad classifications – microglia and macroglia.

The microglia have a defense role as a phagocytic cell. They are found throughout the brain and spinal cord, and can change their shape, especially when they engulf particulate material. They are therefore serving a protective role for the nervous system. Macroglia are subdivided into seven different types, again with each having a specific role.

(1) Astrocytes

These cells fill in the spaces between neurons and provide for structural integrity. They also have processes which join to the capillary blood vessels. These are known as *perivascular end feet*. Therefore, with their close apposition to blood vessels, they are also thought to be responsible for metabolite exchange between the neurons and the vasculature. They are found in the central nervous system.

(2) *Ependymal cells*

There are three types of ependymal cells – ependymocytes, tanycytes and choroidal epithelial cells. The ependymocytes allow for the free movement of molecules between the cerebrospinal fluid (CSF) and the neurons. Tanycytes are generally found in the third ventricle and can be involved in responding to changing hormonal levels of the blood-derived hormones in the CSF. Choroidal epithelial cells are the cells which control the chemical composition of the CSF. They are found in the central nervous system.

(3) Oligodendrocytes

These cells are responsible for the production of myelin sheaths (see next section). They are found in the central nervous system.

(4) Schwann cells

Like oligodendrocytes, Schwann cells are responsible for the production of the myelin sheath (see next section), but in the peripheral nervous system. They also have an additional role in phagocytosis of any debris; therefore help to clean the surrounding environment.

(5) Satellite cells

These cells surround those neurons of the autonomic system and also the sensory system. They maintain a stable chemical balance of the surrounding environment to the neurons. They are therefore found in the peripheral nervous system.

(6) Radial glia

Radial glial cells act as scaffolding onto which new neurons migrate to. They are found in the central nervous system.

(7) Enteric glia

These cells are found within the gastrointestinal tract and aid digestion and maintenance of homeostasis. They are by their very nature found in the peripheral nervous system.

2.1.3 Divisions of the Brain

All vertebrates share the same basic structure to them. The basis of the embryological development is from the neural tube, or the precursor of the central nervous system. At the front, or upper end, of that tube like structure, there are three swellings. These three swellings then become the forebrain, midbrain and hindbrain. The human brain is broadly divided into these three main regions – the forebrain, midbrain and the hindbrain. In mammals, the first part of this neural tube – the forebrain – becomes considerably larger, with the hindbrain remaining rather small in comparison. As previously discussed, the other way to classify the brain based on its components is as follows.

(1) *Telencephalon* (cerebral hemispheres) + *Diencephalon* (thalamus and hypothalamus) = FOREBRAIN

(2) *Mesencephalon* = MIDBRAIN
(3) *Metencephalon* (pons, cerebellum and the trigeminal, abducent, facial and vestibulocochlear nerves) + *Myelencephalon* (medulla oblongata plus the glossopharyngeal, vagus, accessory and hypoglossal nerve nuclei) = HINDBRAIN

The following table provides a broad overview of each of these regions, before they will be discussed in more detail later (Table 2.1).

2.2 FOREBRAIN

2.2.1 Revision Point

As previously discussed, the brain is subdivided as follows:

(1) *Telencephalon* (cerebral hemispheres) + *Diencephalon* (thalamus and hypothalamus) = FOREBRAIN
(2) *Mesencephalon* = MIDBRAIN
(3) *Metencephalon* (pons, cerebellum and the trigeminal, abducent, facial and vestibulocochlear nerves) + *Myelencephalon* (medulla oblongata plus the glossopharyngeal, vagus, accessory and hypoglossal nerve nuclei) = HINDBRAIN

This chapter shall discuss the forebrain. The forebrain is also called the prosencephalon. This is comprised of the smaller portion known as the diencephalon (thalamus and hypothalamus) and the larger portion of the telencephalon (cerebral hemispheres).

2.2.2 Telencephalon

The term telencephalon is really interchangeable with cerebral hemispheres. However, the term cerebrum refers either to the whole brain, or just the forebrain and midbrain.

The cerebral hemispheres occupy a large volume and have several surfaces. Each cerebral hemisphere has a superolateral, inferior and medial surface. The left and right cerebral hemispheres are separated by the *longitudinal fissure*, where the fold of dura passes down called the *falx cerebri*. This partly separates the left and right cerebral hemispheres. The two hemispheres are connected at the lower free edge of the falx cerebri and this connecting bundle of fibers is termed the *corpus callosum*.

Table 2.1. The Functions Associated with the Forebrain, Midbrain and Hindbrain

Region of Forebrain	Functions/Information Processed
Forebrain	
Cerebral hemispheres (TELENCEPHALON)	The cerebral hemispheres process information related to "higher order" functions: Somatosensory information from the opposite side of the body Motor control of the opposite side of the body - Selection and execution of voluntary movements - Movement in space Planning and organization Memory Thought Emotions Problem solving Consciousness Attention Intelligence Language comprehension Processing of information related to vision and sound Speech production and articulation
Thalamus (DIENCEPHALON)	The thalamus can be thought of as a relay and modulating center: Modulation of motor functions (via the ventrolateral and ventroanterior regions of the thalamus) Somatosensory relay Olfactory relay (via the amygdala and prepyriform cortex projections to the mediodorsal thalamus, and on to the frontal lobe, or simply passing through the thalamus without synapse) Visual relay (via the lateral geniculate nucleus) Taste relay (via the ventral posteromedial nucleus) Auditory relay (via the medial geniculate nucleus in the thalamus) Vestibular relay (via the ventral posterolateral nuclei) Thermal relay (via the ventral posterolateral nuclei) Sleep-wake cycle Arousal Consciousness Memory (recollection, familiarization, spatial)
Hypothalamus (DIENCEPHALON)	The hypothalamus has many functions including circadian rhythm, maintenance of body temperature within a narrow effective range, adrenocortical regulation. Broadly speaking, the hypothalamus may be subdivided into two main territories – lateral and medial. Lateral hypothalamus - thirst center, hunger center, predation, reward, motivation - stimulation of parasympathetic outflow - emotions and behavior - descending modulation of spinal neuronal activity Medial hypothalamus - control of the pituitary gland - stimulation of sympathetic outflow - motivational reactions to noxious stimuli - emotions and behavior

(Continued)

Table 2.1. The Functions Associated with the Forebrain, Midbrain and Hindbrain *(cont.)*

Region of Forebrain	Functions/Information Processed
Midbrain	
Mesencephalon	Auditory pathway Vision (the oculomotor and trochlear nerves arise from this point) Autonomic functions Emotional and affective processes Modulation of pain Somatomotor function Autonomic and visceral functions Ascending sensory pathways (spinothalamic, medial lemniscus, trigeminothalamic, auditory)
Hindbrain	
Pons (METENCEPHALON)	Relay center between the cerebral hemispheres and the cerebellum (e.g. medial longitudinal fasciculus and medial lemniscus, spinothalamic, trigeminothalamic, corticobulbar tract, corticospinal tract, rubrospinal tract, tectospinal tract) Location of abducent, facial, trigeminal, superior and lateral vestibular and superior olivatory nuclei, reticular formation and cerebellar peduncles The pons controls a variety of functions – sleep, respiration (via the pontine respiratory group), swallowing, auditory processing, control of the bladder, control of equilibrium, facial expression and sensation
Cerebellum (METENCEPHALON)	Coordination and regulation of motor control (locomotion) Control of balance and posture Vestibular input Cognitive functions (Goldman-Rakic, 1996; Schmahmann and Caplan, 2006) Connections with the hypothalamus for autonomic and emotional functions (Schmahmann and Caplan, 2006)
Trigeminal nerve (METENCEPHALON)	Sensation from face; paranasal sinuses; nose and teeth Muscles of mastication
Abducent nerve (METENCEPHALON)	Innervates the lateral rectus muscle
Facial nerve (METENCEPHALON)	Muscles of facial expression, stylohyoid, stapedius, posterior belly of digastric Parasympathetic innervation of the submandibular and sublingual salivary glands, lacrimal gland and the nasal and palatal glands Anterior two-thirds of the tongue (taste) and palate (general visceral afferent) Concha of the auricle (general sensation)
Vestibulocochlear nerve (METENCEPHALON)	Balance for the vestibular component Hearing for the spiral (cochlear) component
Medulla oblongata (MYELENCEPHALON)	Cardiorespiratory center Reflex centers (e.g. swallowing, vomiting, sneezing, coughing) Vasomotor center

The corpus callosum, which consists of large numbers of nerve fibers interconnect the two hemispheres. Each individual fiber interconnects equivalent points on the two sides. The corpus callosum is shorter than the hemispheres, and has swellings at its rostral and caudal ends (the *genu* and the *splenium*, respectively); this is because the ends must contain not only fibers crossing directly, but also those fibers which interconnect the "extra", protruding parts of the hemisphere.

Each hemisphere has a frontal, temporal and occipital pole moving in an anterior to posterior direction. These are located in the anterior, middle and posterior cranial fossae, respectively. The gray matter of the cerebral cortices is located on its outer aspect, and is thrown into folds called *gyri* (singular, *gyrus*). These are separated by dips in cerebral tissue called *sulci* (singular, *sulcus*).

On the lateral surface the frontal lobe and parietal lobe are separated from one another by the *central sulcus* (which is often not as obvious as its name might suggest). The *pre-central gyrus* is the *primary motor cortex* and many descending motor fibers originate here to descend through the *internal capsule, the cerebral peduncles* and *the pyramids*. There is a map of the body on this region of the brain (i.e. there is *somatotopic localization*) resembling a person hanging upside down over the outer surface of the hemisphere. However, the map is distorted compared with the actual body since it gives predominance to parts over which we have most motor control, e.g. the hand and face. The *post-central gyrus* is the *primary somesthetic (sensory) cortex*. It is to this region that general body sensation (such as touch, proprioception, etc.) is conveyed by fibers, which have come from the thalamus. Again, there is a somatotopic map of the body to permit localization of sensation; again, it is hugely distorted compared with the actual body, with large areas devoted to the face and hand. It is less easy to see the division between the parietal and occipital lobes from the lateral surface.

On the medial surface, the *parieto-occipital sulcus* is found, which serves to separate these two lobes. In addition, the *calcarine sulcus*, which lies in the center of the *primary visual cortex*, begins at the occipital pole, passing anterior to the corpus callosum, at the *splenium*.

The cerebral cortex is divided into frontal, temporal, parietal and occipital lobes. The frontal lobe is bounded by the central and lateral sulci. From this, the parietal lobe passes from the central sulcus to an arbitrary line between the parieto-occipital sulcus and to the *pre-occipital notch*. The occipital lobe is located posterior to that line. The temporal lobes are located anterior to this line, and inferior to the lateral sulcus.

The first of the cranial nerves (olfactory) which, after passing from the nose, enter the skull through the cribriform plate of the ethmoid bone. They end in the olfactory bulbs which are found on the inferior aspect of the frontal lobes. The second of the cranial nerves, the optic nerves, pass through the optic canal of the skull and pass medially to join the one of the opposite side to form the optic chiasma. As the fibers pass posterior, the fibers then separate to form the optic tracts which then pass posterior and around the cerebral peduncles. At this point, these nerves are closely related to the Circle of Willis, which is discussed in more detail in Chapter 6 (Blood Supply of the Brain and Clinical Issues). The infundibular stem of the neurohypophysis, or posterior pituitary, emerges from the tuber cinereum in the interpeduncular fossa. Immediately lateral to the optic tracts, branches of both the anterior and middle cerebral arteries enter here in the area known as the *anterior perforated substance*.

2.2.3 Basal Ganglia

The *basal ganglia* is a mass of gray matter found within the white substance of the cerebral hemispheres. These subcortical regions of gray matter in the forebrain consist of the *caudate nucleus, putamen*, and the *globus pallidus*, as well as the *subthalamic nucleus* and the *substantia nigra*. Information from the cerebral cortex is fed to the caudate and putamen, and from there to the globus pallidus, before returning via the thalamus to motor areas of the cortex. The role of the basal ganglia in movement is only poorly understood, but basal ganglia disorders often have profound effects on movement. Disorders which can affect the basal ganglia are briefly mentioned below.

(1) *Parkinson's disease* which is associated with a loss of cells in the *substantia nigra* which project to the caudate/putamen and use *dopamine* as their neurotransmitter. Treatments include administering *L-Dopa* (the precursor of dopamine) or implanting dopaminergic neurons or stem cells. Symptoms of Parkinson's disease can include
    (a) difficulty in initiating and terminating movement
    (b) tremor at rest
    (c) slowness and impoverishment of movement, loss of "automatic" movements
    (d) tiny writing, deadpan face, soft monotonous speech, and shuffling gait

(2) Hemiballismus (violent flailing of the limbs/trunk)
(3) Huntington's chorea, which is a fatal condition, involves
    degeneration of the basal ganglia before becoming more widespread.

2.2.4 Corpus Striatum

The corpus striatum is made up of the caudate nucleus and the lentiform nucleus. The caudate nucleus bulges into the lateral ventricle and is comprised of a head, body and a tail. The caudate nucleus is an arched structure and frequently can appear twice on sectioning the brain. The head is anterior posterior to the genu of the corpus callosum. The body of the caudate nucleus extends posterior and lateral to the position of the thalamus. The tail of the caudate nucleus curves in an inferior and anterior direction into the temporal lobe ending in the amygdaloid body.

The lentiform nucleus is found lateral to the head of the caudate nucleus and thalamus. Anteriorly it is connected to the head of the caudate nucleus by gray matter. The lateral aspect of the lentiform nucleus is referred to as the putamen, and is closely related to the claustrum and the insula. The two medial parts of the lentiform nucleus are referred to as the globus pallidus.

2.2.5 Internal Capsule

The internal capsule is a large territory of white matter found between the following structures.

(1) Lentiform nucleus (laterally)
(2) Caudate nucleus (head of) and the thalamus (medially)

The internal capsule is comprised of an anterior limb (found between the caudate and lentiform nuclei), genu (V-shaped bend), posterior limb (between the thalamus and the lentiform nucleus), and retrolentiform and sublentiform divisions.

When the fibers are followed superiorly, they fan out into the cerebral hemispheres, and this point is called the corona radiata. These fibers of the corona radiata are intersected by the fibers of the corpus callosum.

As well as carrying information past the basal ganglia, the internal capsule also contains both ascending and descending tracts. A large proportion of the internal capsule is comprised of the corticospinal tract, which transmits information from the primary motor cortex inferiorly to terminate on the lower motor neurons found in the ventral horn of the spinal cord. Further details of the corticospinal tract are found in Chapter 9.

The clinical significance of the internal capsule lies in the fact that the *lenticulostriate arteries* supply a large portion of it. Occlusion or rupture of these vessels will result in interruption of a large number of fibers that pass through here, and can result in disastrous consequences for the patient as it not only can result in hemiplegia or hemiparesis, but also can result in a large sensory deficit for the patient. This is discussed in detail in Chapter 6.

2.2.6 Diencephalon
The diencephalon is comprised of several components – the thalami, the lateral and medial geniculate bodies and finally the hypothalamus. All of these structures surround part of the ventricular system circulating the cerebrospinal fluid (see later).

2.2.6.1 Thalamus
The thalamus is divided up according to its nuclei – anterior, lateral, medial, midline and intralaminar. Similarly, the thalamus can be classified according to the functional territories – specific, non-specific and association nuclei.

2.2.6.1.1 Specific Thalamic Nuclei
- Input – from sensory pathways, e.g. spinothalamic pathway, retina (vision), inferior colliculus (hearing) and the trigeminothalamic tract (somatosensory from the face)
- Output – specific regions of the cerebral cortex dealing with the projection areas. These would project to the auditory, visual, somesthetic and temporal neortex regions

2.2.6.1.2 Non-specific Thalamic Nuclei
- Input – specific and association nuclei and the reticular formation
- Output – layer I of the neocortex (non-specifically) and specific thalamic nuclei

2.2.6.1.3 Association Nuclei
- Input – basal ganglia, limbic nuclei, non-specific thalamic nuclei and the cerebral cortex
- Output – wide and varied

In terms of the functional organization of the thalamus, one area that has received a great deal of attention has been the mediodorsal thalamus (MDT). It is a major component of the thalamus of all mammals, and is especially developed in humans (Le Gros Clark, 1932a,b). Indeed the functions of the MDT are wide and varied with it being well established that it plays a major role in emotional changes, anterograde amnesia where postoperative learning is severely affected especially in visual memory tasks (Schulman, 1957; Zola-Morgan and Squire, 1985; Parker et al., 1997; Gaffan and Watkins, 1991; Gaffan and Parker, 2000; Mitchell et al., 2007) and memory acquisition (Mitchell and Gaffan, 2008).

Major reciprocal connections exist between the MDT and the prefrontal cortex (Parker and Gaffan, 1998; McFarland and Haber, 2002; Erickson and Lewis, 2004) and (along with the orbitofrontal cortex) has descending pathways linked to the amygdala and the hypothalamic nuclei (Barbas et al., 2003) in the primate. As the MDT has been shown to serve a role in nociceptive processing (Casey, 1966; Palestini et al., 1987; Dostrovsky and Guilbaud, 1990), it could be said that a loop exists forming the "basolateral limbic system" composed of the MDT, amygdala, orbitofrontal cortex including that also of the rat (Krettek and Price, 1977; Sarter and Markowitsch, 1983, 1984; Bachevalier and Mishkin, 1986; Cassell and Wright, 1986; Gaffan et al., 1993) and the prefrontal cortex (Fuster, 1997). Therefore, the MDT could serve a role in the motivational and affective components of pain, including that of visceroception and visceronociception. As well as the thalamus playing a major role in nociceptive processing, in recent years an increasing number of studies have been undertaken in examining the role of the medulla oblongata in its role in nociception. The functional organization of the thalamus and its projections will be dealt with in more detail in Chapter 3.

2.2.6.2 Hypothalamus

The hypothalamus has several key territories called the ventromedial, paraventricular, suprachiasmatic, supraoptic, arcuate, dorsomedial and tuberal nuclei, as well as the preoptic region and mammillary bodies. It is responsible for autonomic regulation and visceral behavior. It is also crucial to endocrine function and essential in temperature regulation, feeding, drinking and the control of water within the body. It is also related to rage and aggression and flight behavior.

2.2.7 Ventricles

The ventricles of the brain circulate cerebrospinal fluid. There are two *lateral ventricles* and a *third* and *fourth ventricle*. These will each be described from a superior to inferior direction. The lateral ventricles (left and right) communicate with the third ventricle below via an *interventricular foramen* on each side. The third ventricle communicates with the fourth ventricle below it via the *cerebral aqueduct*. The fourth ventricle then becomes continuous with the *central canal* of the medulla and the spinal cord, and opens by apertures into the *subarachnoid space*.

Within the ventricles and also the central canal of the spinal cord, specialized neuroglia called *ependymal cells* produce the cerebrospinal fluid, and line these territories. Within the ventricles, vascular fringes of pia mater invaginate their covering of the ependymal cells and project into the ventricles. This combination of the folds of vascular pia mater and ependymal cells creates the *choroid plexus*, and it passes into the cavities of the lateral, third and fourth ventricles. It is the choroid plexus that produces the cerebrospinal fluid.

2.2.7.1 Lateral Ventricle

There are two lateral ventricles on each side of the cerebral hemispheres. They communicate inferiorly with the third ventricle through the *interventricular foramen*. The area of the lateral ventricle anterior to the foramen is the *anterior horn*, or *first part*. Posterior to this is the *central part*. The front, middle and back portions of the central part are numbered second, third and fourth parts, respectively. The fourth part of the ventricle divides into the fifth part, also called the *posterior horn*, and the sixth part is referred to as the *inferior horn*. The anterior, inferior and posterior horns are located in the frontal, temporal and occipital lobes, respectively.

- Anterior horn – bounded below by the rostrum, in front by the genu, and superiorly by the trunk of the corpus callosum
  - Limited laterally by the caudate nucleus
  - The septum pallucidum separates one lateral ventricle from its neighbor
- Central part – lies beneath the trunk of the corpus callosum
  - Found on the thalamus and body of the caudate nucleus
  - Separated from each other by the posterior portion of the septum pallucidum

- Posterior horn – asymmetrical
  - Bounded by the *tapetum* on the lateral side
  - Elevations may be found laterally (upper and lower)
- Inferior horn – bounded laterally by the tapetum
  - Inferiorly is the hippocampus
- Blood supply of choroid plexus – anterior choroid artery (from the internal carotid artery)
  - Posterior choroid artery (from the posterior cerebral artery)

2.2.7.2 Third Ventricle

This is a narrow region between the two thalami. The thalami are adherent to one and other over a variable area and this is the *interthalamic adhesion*. The floor of the third ventricle is formed by the hypothalamus and is crossed anteriorly by the optic chiasma.

The third ventricle communicates with the lateral ventricles by the *interventricular foramina*. Each of these foramina is found anterosuperiorly, toward the anterior limit of the thalamus.

- Blood supply – posterior choroid artery (from the posterior cerebral artery)

2.2.7.3 Fourth Ventricle

The fourth ventricle is found in the posterior region of the pons and medulla and is rhomboid in shape. Superiorly, it narrows to become continuous with the aqueduct of the midbrain. Inferiorly, it narrows and leads into the central canal of the medulla. This in turn is continuous with the central canal of the spinal cord. The fourth ventricle is widened at the point called the lateral recess.

The anterior boundary, or floor is formed by the pons superiorly and medulla inferiorly. The nuclei of origin of the vestibulocochlear nerves are closely related to this. The median groove divides the floor into left and right halves. Each half is divided by the *sulcus limitans* into the *medial*, or *basal portion* and the *lateral*, or *alar portion*.

The lowermost portion of the floor of the fourth ventricle is called the *calamus scriptorius*, as it appears to resemble the tip of a pen. This region contains the cardiorespiratory, deglutition and vasomotor centers.

The posterior boundary or roof of the fourth ventricle is very thin and concealed by the cerebellum. It consists of white matter referred to as the superior and inferior medullary vela. This is lined by ependymal. There is a deficiency in the lower portion of the roof called the *median aperture (Foramen of Magendie)*. At the median aperture, there is direct communication with the subarachnoid space. The ends of the lateral recess have openings called the *lateral apertures (Foramen of Luschka)*. It is through the median and lateral apertures that the cerebrospinal fluid enters the subarachnoid space. The blood supply of the choroid plexus is from the cerebellar branches of the basilar and vertebral arteries.

2.2.8 Cerebrospinal Fluid

The cerebrospinal fluid (CSF) is a clear fluid which is produced by the choroid plexus and circulates around the brain and spinal cord. The following presents the key facts about CSF.

(1) Total volume of 100–150ml at any one time
(2) Approximately 500ml is produced in a 24h period
(3) Turnover of CSF is up to fourfold each day
(4) Ependymal cells produce approximately two-thirds of the circulating CSF
(5) The remainder of the CSF is produced by the ventricular surfaces and the lining of the subarachnoid space
(6) CSF has a higher concentration of sodium and chloride, but less protein, glucose, potassium and calcium than blood

2.2.8.1 Functions of CSF

(1) Certain degree of protection, but not in any type of high impact injury. It will "slow" movement of the brain in quick rotations of the head but not necessarily protect against significant injury
(2) Stabilization of the chemical balance of the CSF and allowing clearance of waste products into the bloodstream
(3) Maintenance of buoyancy of the brain. As the brain weighs approximately 1.5 kg, the surrounding fluid reduces the net weight of the brain within the cranial cavity

A sample of cerebrospinal fluid can be obtained during a lumbar puncture and this is discussed further in Chapter 7.

2.3 PATHOLOGIES

(1) *Dysphasia* defects in the understanding and construction of speech
(2) *Dysarthria and dysphonia* defects in its mechanical production

2.3.1 Dysphasia

The neural substrate for human speech is located in the *left cerebral hemisphere* in some 97% of the population. Two areas are of importance are as follows.

(1) A sensory speech area located near the auditory cortex and covering quite a large area of the temporal and parietal lobes. It has an important core region – *Wernicke's area.*
(2) A motor speech area located in the frontal lobe near the face, tongue and larynx regions of the motor cortex – *Broca's area.*

Damage to Wernicke's or Broca's area (or pathways that link them) leads to dysphasia. Such damage might result from a stroke or a tumor, and might occasionally be widespread enough to affect both areas. If only one area is affected, the results are different depending which of the two is involved.

2.3.1.1 If Wernicke's Area is Damaged

(1) The speech of others is heard (therefore the patient is not deaf) but is not understood. Conversion of sounds to ideas does not occur, and the patient cannot obey instructions
(2) Patients cannot monitor their own speech, which becomes excessive, unintelligible, full of jargon and new made-up words (neologisms).

Patients usually have no insight into their state of deficiency, and may appear so withdrawn from reality that diagnoses of other mental disorders are made.

Two other conditions also carry Wernicke's name, who founded the term Wernicke's area, but not related to that area so this confusion should be avoided. Initially Werniceks' area was discovered shortly after Paul Broca had discovered language related areas within the frontal lobe. Wernicke had discovered that not all deficits with language (in those patients with these issues) were related to Broca's area, but rather they were related to a portion of the brain to the left superior temporal gyrus on its posterior aspect.

Other conditions with Wernicke's name, but not related to Wernicke's area are worth a mention here to avoid any confusion specifically with Wernicke's area.

2.3.1.1.1 Wernicke's encephalopathy

This condition is also now referred to as Wernicke's disease. It results from the deficiency of the B vitamin, particularly thiamine, and is generally found in alcoholic patients. Patients affected by Wernicke's encephalopathy have a triad of nystagmus, ataxia and ophthalmoplegia (most commonly of the lateral recti).

Wernicke's encephalopathy can present with this triad, though not always. It can also present with headaches, anorexia and vomiting, as well as confusion. The symptoms really depend on the brain region involved as it can affect a variety of areas including the medulla oblongata, hypothalamus, brain stem tegmentum or affect the cerebral cortex more globally and diffusely.

Wernicke's encephalopathy is investigated by history taking; clinical examination and it may also be noted that the patient has depleted levels of red cell transketolase and thiamine or a raised plasma pyruvate. In addition, other vitamin and mineral levels should also be examined in these patients. Refer to local guidelines for further investigation of this condition.

The treatment of Wernicke's encephalopathy is by intravenous or intramuscular thiamine injections. If these deficiencies are treated early, many of the symptoms will improve. However, the underlying issues have to be investigated, including that of potentially alcoholism. Specialist input and advice will be necessary for these patients.

2.3.1.1.2 Korsakoff's syndrome

Korsakoff's syndrome is also due to a thiamine deficiency, and therefore, can be linked to Wernicke's encephalopathy. If both conditions are present it is referred to as Wernicke–Korsakoff syndrome.

Korsakoff syndrome affects the memory of patient and they exhibit several features. They will present with amnesia both of forming new memories (anterograde amnesia) and remembering old memories that have happened in their lifetime (retrograde amnesia). As well as difficulty with the formation of memories and remembering the past, this can be distressing, and as such, these patients will create stories to fit into those areas where they are unable to recollect. This is referred to as *confabulation*. In addition, these patients will have difficulty joining in with conversations, and will not have insight into the world around them. The patient will also lose interest in the world around them and have *apathy*.

2.3.1.1.3 Wernicke–Korsakoff syndrome

There is a close association between these two syndromes and are a result of thiamine deficiency. If not identified early on and treated with thiamine, this condition can result in either institutionalization or death. It can also lead to permanent irreversible brain damage. It is associated with an acute or gradual encephalopathy and generally is associated with alcohol abuse. It comprises a mixture of signs and symptoms typically found with each of these conditions mentioned above. Therefore, as well as the triad of nystagmus, ataxia and ophthalmoplegia (most commonly of the lateral recti), these patients will also have a confusional state and problems with their current and past memory, alongside confabulation.

Wernicke–Korsakoff syndrome typically affects the frontal lobe, thalamus, mammillary bodies and also the periaqueductal gray matter.

This syndrome is generally diagnosed from clinical history and a comprehensive neurological examination. It may be relevant to consult with the neurological team to aid the diagnosing, but if suspected, thiamine can be given if the patient presents acutely, perhaps through accident and emergency/emergency room.

2.3.1.2 If Broca's Area is Damaged

(1) The patient understands the speech of others and spoken words lead to sensible ideas. However,
(2) Patients cannot find words to express themselves, are non-fluent, hesitant, make grammatical errors, omissions, and may speak like telegrams. They may stammer and, in extreme cases, be rendered mute.

Patients usually have insight into the deficit and become very frustrated.

Because of the way in which damage to the brain occurs, it is unlikely to be restricted to the speech area alone. Damage to the region that includes Wernicke's area may, e.g. also result in difficulties with writing (dyslexia) and numbers, visual impairment (part of the visual pathway, called Meyer's loop, runs through the temporal lobe) and impaired memory.

Damage to the region that includes Broca's area might also result in weakness of the R. face and upper limb (and even sensory loss from these regions) by involving the sensorimotor strip.

Dysphasias may also occur because of damage which disconnects links between Wernicke's and Broca's areas. Do not assume that this is well understood.

2.3.1.3 "Split-brain" Patients
A few patients have had the corpus callosum surgically sectioned (usually in cases of severe epilepsy). This disconnects the two hemispheres. An object held in the right hand can be named; an object held in the left hand cannot, even though it can be used correctly (a key, a pencil).

2.3.1.4 Removing the Left Hemisphere
The most radical neurosurgical procedure may involve removing a hemisphere completely (hemispherectomy) or disconnecting it from the rest of the brain. It may be done for large tumors, or for the prevention of debilitating seizures. Where operations on the left hemisphere are performed during childhood, it is sometimes possible for speech function to establish itself in the right hemisphere. There is debate on the "critical age" beyond which it is impossible to transfer, but 9–10 seems an upper limit. Removal of the left hemisphere in an adult leads to loss of speech (aphasia).

2.3.2 Dysarthria
Dysarthiras are mechanical defects in speech, arising from difficulties in moving lips, tongue, palate, pharynx, larynx, etc. In other words, clear pronunciation is impaired due to a neuromuscular deficit, but the "sense" of speech is quite normal.

2.3.2.1 Pseudobulbar Palsy
This involves damage to pathways connecting the cerebral cortex to cranial nerves in the brain stem. These cranial nerves are the motor

nerves V (muscles of mastication), VII (facial musculature), IX, X, XI (pharynx and larynx), XII (tongue). Unilateral damage does little to affect speech unless VII is involved, when some slurring occurs. Bilateral damage does disturb speech; it is most likely to result from some global effect on nerves, such as multiple sclerosis, motor neuron disease, etc.

2.3.2.2 Bulbar Palsy

It usually involves damage to the brain stem itself (and hence to the cranial nerves referred to above which are located within the brain stem). However, it is a loose term sometimes applied to damage to the nerve axons or neuromuscular junctions, or to the muscles they supply. In any event, speech defects would be the least of worries; there would be difficulty in swallowing (leading to aspiration of food and fluids and pneumonia) and, probably, difficulties in respiration. Bulbar palsy is life threatening.

Dysarthrias also occur because of damage to other brain regions which control movement, such as the basal ganglia or the cerebellum. Damage to the basal ganglia may result in quiet and unemotional (deadpan) speech, while cerebellar damage may result in "drunken" speech.

2.3.3 Dysphonia

Dysphonia is quiet or hoarse speech resulting from abnormalities in the passage of air across the vocal folds. Possible causes may be as follows:

(1) infections or pathological changes such as laryngitis, the development of nodules on the vocal folds, or carcinoma of the larynx;
(2) damage to the innervation of the laryngeal muscles; and
(3) weakness of the chest muscles.

2.4 MIDBRAIN

The following is a summary of the main components of the brain. The midbrain is also referred to as the mesencephalon.

(1) *Telencephalon* (cerebral hemispheres) + *Diencephalon* (thalamus and hypothalamus) = FOREBRAIN
(2) *Mesencephalon* = MIDBRAIN
(3) *Metencephalon* (pons, cerebellum and the trigeminal, abducent, facial and vestibulocochlear nerves) + *Myelencephalon* (medulla oblongata plus the glossopharyngeal, vagus, accessory and hypoglossal nerve nuclei) = HINDBRAIN

The midbrain is comprised of the cerebral peduncles, tectum, tegmentum, cerebral aqueduct and some cranial nerve nuclei. The following table summarizes the main functions of each of those divisions (Table 2.2).

Further detail of the midbrain and its structures can be found in Chapter 4.

2.5 HINDBRAIN

The following is a summary of the main regions of the brain, detailing what comprises the hindbrain (Table 2.3).

(1) *Telencephalon* (cerebral hemispheres) + *Diencephalon* (thalamus and hypothalamus) = FOREBRAIN
(2) *Mesencephalon* = MIDBRAIN
(3) *Metencephalon* (pons, cerebellum and the trigeminal, abducent, facial and vestibulocochlear nerves) + *Myelencephalon* (medulla oblongata, plus the glossopharyngeal, vagus, accessory and hypoglossal nerve nuclei) = HINDBRAIN

Table 2.2. The Key Functions of Each of the Main Regions Within the Midbrain	
Region of Midbrain	**Functions/Information Processed**
Cerebral peduncles	Motor functions
Tectum	Auditory information transmission Influence of spinal motor neurons in the cervical region Visual reflexes Movement of the head and eyes (voluntary and involuntary)
Tegmentum	Transmission of sensory information Motor functions, e.g. of limbs and voluntary actions Arousal
Cerebral aqueduct	Transmission of cerebrospinal fluid, connecting the third ventricle above to the fourth ventricle below

Table 2.3. The Functions Associated With Each of the Regions of the Hindbrain

Region of Hindbrain	Functions/Information Processed
Metencephalon	
Pons	Transmission of tract pathways from the cerebral cortex to the cerebellum and medulla, as well as somatosensory information to the thalamus Sleep Respiration (pneumotaxic center) Equilibrium and hearing Sensation of the face Eye movements Posture Deglutition and taste
Cerebellum	Coordination of motor activities
Trigeminal nerve	Sensation from face; paranasal sinuses; nose and teeth Muscles of mastication
Abducent nerve	Innervates the lateral rectus muscle
Facial nerve	Muscles of facial expression, stylohyoid, stapedius, posterior belly of digastric Parasympathetic innervation of the submandibular and sublingual salivary glands, lacrimal gland and the nasal and palatal glands Anterior two-thirds of the tongue (and palate) Concha of the auricle
Vestibulocochlear nerve	Balance for the vestibular component; hearing for the spiral (cochlear) component
Myelencephalon	
Medulla oblongata	Autonomic related functions including respiration and regulation of the cardiac center. Deglutition, vomiting, sneezing and coughing
Glossopharyngeal nerve	Stylopharyngeus Parotid gland for parasympathetic innervation Taste from the posterior one-third of the tongue External ear Pharynx; parotid gland; middle ear; carotid sinus and body
Vagus nerve	Pharyngeal constrictors; laryngeal muscles (intrinsic); palatal muscles; upper two-thirds of esophagus; heart; trachea and bronchi; gastrointestinal tract Taste from the palate and the epiglottis Auricle; external auditory meatus; posterior cranial fossa dura mater Gastrointestinal tract (to last one-third of the transverse colon); pharynx and larynx; trachea and bronchi; heart
Accessory nerve	Innervates the sternocleidomastoid and trapezius muscles
Hypoglossal nerve	Extrinsic and intrinsic muscles of the tongue. Palatoglossus is not supplied by the hypoglossal nerve. It is supplied by the glossopharyngeal nerve

REFERENCES

Azevedo, F.A., Carvalho, L.R., Grinberg, L.T., Farfel, J.M., Ferretti, R.E., Leite, R.E., Jacob Filho, W., Lent, R., Herculano-Houzel, S., 2009. Equal numbers of neuronal and nonneuronal cells make the human brain an isometrically scaled-up primate brain. J. Comp. Neurol. 513 (5), 532–541.

Bachevalier, J., Mishkin, M., 1986. Visual recognition impairment follows ventromedial but not dorsolateral prefrontal lesions in monkey. Behav. Brain Res. 20, 249–261.

Barbas, H., Saha, S., Rempel-Clower, N., Ghashghaei, T., 2003. Serial pathways from primate prefrontal cortex to autonomic areas may influence emotional expression. BMC Neurosci. 4, 25.

Casey, K.L., 1966. Unit analysis of nociceptive mechanisms in the thalamus of the awake squirrel monkey. J. Neurophysiol. 29, 727–750.

Cassell, M.D., Wright, D.J., 1986. Topography of projections from the medial prefrontal cortex to the amygdala in the rat. Brain Res. Bull. 17, 321–333.

Dostrovsky, J.O., Guilbaud, G., 1990. Nociceptive responses in medial thalamus of the normal and arthritic rat. Pain 40, 93–104.

Erickson, S.L., Lewis, D.A., 2004. Cortical connections of the lateral mediodorsal thalamus in cynomolgus monkeys. J. Comp. Neurol. 473, 107–127.

Fuster, J.M., 1997. The Prefrontal Cortex: Anatomy, Physiology, and Neuropsychology of the Frontal Lobe. Raven Press, New York.

Gaffan, D., Parker, A., 2000. Mediodorsal thalamic function in scene memory in rhesus monkeys. Brain 123, 816–827.

Gaffan, D., Watkins, S., 1991. Mediodorsal thalamic lesions impair long-term visual associative memory in macaques. Eur. J. Neurosci. 3, 615–620.

Gaffan, D., Murray, E.A., Fabre-Thorpe, M., 1993. Interaction of the amygdala with the frontal lobe in reward memory. Eur. J. Neurosci. 5, 968–975.

Goldman-Rakic, P., 1996. Regional and cellular fractionation of working memory. Proc. Natl. Acad. Sci. U.S.A. 93, 13473–13480.

Krettek, J.E., Price, J.L., 1977. The cortical projections of the mediodorsal nucleus and adjacent thalamic nuclei in the rat. J. Comp. Neurol. 171, 157–192.

Le Gros Clark, W.E., 1932a. The structure and connections of the thalamus. Brain 55, 406–470.

Le Gros Clark, W.E., 1932b. An experimental study of thalamic connections in the rat. Philos. Trans. R. Soc. London [Biol.] 221, 1–28.

McFarland, N.R., Haber, S.N., 2002. Thalamic relay nuclei of the basal ganglia form both reciprocal and nonreciprocal cortical connections, linking multiple frontal cortical areas. J. Neurosci. 22, 8117–8132.

Mitchell, A.S., Gaffan, D., 2008. The magnocellular mediodorsal thalamus is necessary for memory acquisition, but not retrieval. J. Neurosci. 28, 258–263.

Mitchell, A.S., Baxter, M.G., Gaffan, D., 2007. Dissociable performance on scene learning and strategy implementation after lesions to magnocellular mediodorsal thalamic nucleus. J. Neurosci. 27, 11888–11895.

NIH, 2014. US National Library of Medicine. An ancient medical treasure at your fingertips. <http://www.nlm.nih.gov/news/turn_page_egyptian.html> (accessed 22.04.2014.).

Palestini, M., Mariotti, M., Velasco, J.M., Formenti, A., Mancia, M., 1987. Medialis dorsalis thalamic unitary response to tooth pulp stimulation and its conditioning by brainstem and limbic activation. Neurosci. Lett. 78, 161–165.

Parker, A., Gaffan, D., 1998. Interaction of frontal and perirhinal cortices in visual object recognition memory in monkeys. Eur. J. Neurosci. 10, 3044–3057.

Parker, A., Eacott, M.J., Gaffan, D., 1997. The recognition memory deficit caused by mediodorsal thalamic lesion in non-human primates: a comparison with rhinal cortex lesion. Eur. J. Neurosci. 9, 2423–2431.

Sarter, M., Markowitsch, H.J., 1983. Convergence of basolateral amygdaloid and mediodorsal thalamic projections in different areas of the frontal cortex in the rat. Brain Res. Bull 10, 607–622.

Schmahmann, J.D., Caplan, D., 2006. Cognition, emotion and the cerebellum. Brain 129 (2), 290–292.

Schulman, S., 1957. Bilateral symmetrical degeneration of the thalamus: a clinico-pathological study. J Neuropathol. Exp. Neurol. 16, 446–470.

Zola-Morgan, S., Squire, L.R., 1985. Amnesia in monkeys after lesions of the mediodorsal nucleus of the thalamus. Ann. Neurol. 17, 558–566.

CHAPTER 3

Forebrain

3.1 INTRODUCTION

The forebrain is comprised of the telencephalon (cerebral hemispheres) and the diencephalon (thalamus and hypothalamus) as previously discussed. This chapter will provide summary tables as to the key functions of each of these territories. First, the telencephalon will be dealt with which will break down the cerebral hemispheres into the lobes (frontal, parietal, temporal and occipital) providing key information about the various regions and associated functions. The subdivisions are classified according to Brodmann's classification of the areas. This was a wide study which was based on the cytoarchitecture as determined by the Nissl staining method of all of the cerebral cortex (Brodmann, 1909). This text was in German, but a translation was developed for this and edited by Garey (2006). The key areas and their functions are detailed below. In addition, the basal ganglia and the limbic system will be dealt with in the same tabulated format. The structures of the diencephalon will be dealt with in a similar manner.

3.2 TELENCEPHALON

3.2.1 Cerebral Cortex

The table below will provide an overview of the subdivisions of the frontal, parietal, temporal and occipital lobes and describes the functions of each of these areas (Table 3.1).

Table 3.1. The Subdivisions of the Frontal, Parietal, Temporal and Occipital Lobes and Describes the Functions of Each of These Areas

Cerebral Cortex		
Lobe	Subdivision	Functions
Frontal lobe	Primary motor cortex (area 4)	Execution and regulation of movement of the opposite side of the body
	Premotor cortex (area 6)	Movement which requires visual input Coordination of voluntary movements
	Frontal eye fields (area 8)	Coordination of voluntary movements of the eyes

(Continued)

Essential Clinical Anatomy of the Nervous System. http://dx.doi.org/10.1016/B978-0-12-802030-2.00003-0

Table 3.1. The Subdivisions of the Frontal, Parietal, Temporal and Occipital Lobes and Describes the Functions of Each of These Areas *(cont.)*

Cerebral Cortex		
Lobe	Subdivision	Functions
	Dorsolateral prefrontal cortex (area 9, 46)	Memory Planning and execution Decision making
	Prefrontal cortex (area 10)	Regulation of cognition, emotions and visceral functions
	Orbitofrontal area (area 11, 12)	Comparison of reward and punishment in a given situation
	Anterior cingulate gyrus (area 24)	Attention Motivation Regulation of emotional response Recognition of error
	Ventromedial prefrontal cortex (area 25)	Regulation of emotional response
	Anterior cingulate cortex (area 32,33)	Attention Motivation Regulation of emotional response Recognition of error
	Broca's area (area 44, 45)	Language production and comprehension i.e. motor expression of speech
Parietal lobe	Primary somesthetic cortex (areas 1-3)	Sensory input for pain and tactile information Conscious proprioception Position sense
	Posterior parietal cortex (areas 5 and 7)	Appropriate movements for given motor task Integration with the somatosensory system and visual input to the premotor cortex (area 6)
	Posterior cingulate cortex (area 23)	Integration center Cognition Regulation of emotion
	Cingulate cortex (area 26, 29, 30)	Memory recall Emotions Learning Modulation of functions of the hypothalamus
Temporal lobe	Inferior, middle and superior temporal gyri (areas 20, 21 and 22, respectively)	Processing of visual information (e.g. recognition and perception)
	Wernicke's area (area 22)	Interpretation of speech and the written language
	Fusiform gyrus (area 37)	Recognition e.g. face and shapes
	Primary auditory cortex (area 41, 42 and 22)	Processing of auditory information
Occipital lobe	Primary visual cortex (V1; area 17)	Processing of visual information Pattern recognition Detection of static and moving objects
	Secondary visual cortex (V2; area 18)	Visual association
	Tertiary visual cortex (V3 and V5; area 19)	Visual form (V3) and detection of motion (V5)

3.2.2 Basal Ganglia

The basal ganglia are subcortical structures which are responsible for the regulation of motor functions with complex input and output pathways. The basal ganglia is comprised of the neostriatum (putamen and caudate nucleus), paleostriatum (globus pallidus) and the subthalamic nucleus and substantia nigra. The afferent input to the basal ganglia is primarily via the neostriatum (i.e. the putamen and caudate nucleus). The major output of the basal ganglia is from the medial pallidus. The table below will provide an overview as to the input/output and function of each of these regions (Table 3.2).

Table 3.2. The Input, Output and Functions of Each of the Specific Regions of the Basal Ganglia

Basal Ganglia				
Component	**Specific Territory**	**Input**	**Output**	**Functions**
Neostriatum	Putamen	Input from the primary and secondary motor cortices Input from the primary somesthetic cortex Input from the substantia nigra (pars compacta) exhibiting both excitatory and inhibitory effects via dopamine*	Inhibitory effect to both the lateral and medial globus pallidus	Motor functions
	Caudate nucleus	Input from frontal eye fields, limbic areas of the cortex and cortical association regions (parietal) Input from the substantia nigra exhibiting both excitatory and inhibitory effects via dopamine*		Cognitive aspects of emotion and movement (including that of the eye)
Paleostriatum	Globus pallidus	Inhibitory input from the neostriatum to the lateral and medial globus pallidus Excitatory input from the subthalamic nucleus to the medial globus pallidus	Inhibitory input to the thalamus (ventrolateral, ventral anterior and centromedian nuclei) from the medial globus pallidus Inhibitory input from the lateral globus pallidus to the subthalamic nucleus	Inhibitory role balancing the excitatory effect of the cerebellum Movement occurring subconsciously

(Continued)

Table 3.2. The Input, Output and Functions of Each of the Specific Regions of the Basal Ganglia *(cont.)*

Basal Ganglia				
Component	Specific Territory	Input	Output	Functions
Subthalamic nucleus	Inhibitory input from the lateral globus pallidus (via GABA)	Excitatory effect on the medial globus pallidus (via glutamate)		
Substantia nigra	Pars reticulata	Inhibitory influence (via GABA) from the neostriatum	Inhibitory output (via GABA) to the superior colliculus and the ventrolateral and ventral anterior thalamic nuclei	Control of eye movements and orientation
	Pars compacta	Inhibitory input from the pars reticulata	Excitatory (D2 receptors) and inhibitory (D1 receptors) input to the neostriatum via dopamine	Prediction of reward with behavior Motor control

*Dopamine acting through D1 receptors is excitatory to inhibitory GABAergic neurons. Dopamine acting through the D2 receptors inhibits GABAergic neurons.

3.3 CLINICAL ASSESSMENT

Assessment of the basal ganglia can be undertaken by doing a full neurological examination, and assessing motor function, as described in Chapter 8.

3.4 PATHOLOGIES

3.4.1 Parkinson's Disease

Parkinson's disease is associated with a loss of cells in the *substantia nigra* which project to the caudate/putamen and use *dopamine* as their neurotransmitter. It presents clinically with a wide range of signs and symptoms resulting from the depletion of dopamine as follows.

(a) *Tremor* – "Pill rolling" where there is rhythmic contraction of the thumb and fingers
(b) *Rigidity* – *lead pipe rigidity* where the limbs will resist passive extension through movement. Also, *cogwheel rigidity* is present where the patient has combined rigidity and tremor
(c) *Bradykinesis* – slow in initiating movement, e.g. walking,
(d) *Mask like face* – expressionless face
(e) *Shuffling gait* with a flexed trunk. When walking the patient with Parkinson's disease finds it difficult to stop.

Parkinson's disease is a main cause of *parkinsonism* which is the term used for the collection of symptoms related to motor control. Symptoms start at the age of 60–70 and are progressive and can lead to dementia. Research recently out has shown that depression and anxiety are twice as common in those with Parkinson's disease (de la Riva et al., 2014). The cause of Parkinson's disease is still not known.

Treatment of Parkinson's disease is by using L-dopa and dopamine agonists. In the early stages of the disease, they can be effective, but as the disease progresses, and there is continual loss of dopaminergic neurons, the effectiveness of the drugs reduce.

The clinician should be aware of depression and also advise a balanced diet to reduce any effects of the nervous system on the gastrointestinal tract.

3.4.2 Limbic System

The word limbic literally means border. These are a collection of structures which used to be thought of as being responsible for olfaction. The main structures which comprise the limbic system are the hippocampal formation, amygdala, septal area, anterior cingulate gyrus and related regions of the cortex. The limbic system is involved in a variety of functions including long-term memory, emotions, behavior and olfaction. The following table (Table 3.3) highlights the main regions mentioned, the afferent (input) and efferent (output) information passing in or leaving to, and the functions associated with each of the divisions.

Table 3.3. The Input, Output and Functions of Each of the Specific Regions of the Limbic System

Limbic System				
Component	Specific Territory	Input	Output	Functions
Hippocampal formation	Hippocampus	Entorhinal cortex Diagonal band of broca Anterior cingulate gyrus Premammilary region Prefrontal cortex Brainstem reticular formation	Septal area Medial hypothalamus Anterior thalamic nucleus Mamillary bodies Cingulate cortex Entorhinal cortex Prefrontal cortex Contralateral hippocampus	Learning Memory Endocrine functions Autonomic activity Behavior Navigation and spatial awareness

(Continued)

Table 3.3. The Input, Output and Functions of Each of the Specific Regions of the Limbic System *(cont.)*

		Limbic System		
	Dentate gyrus	Entorhinal cortex	Interneurons	Creation of memory
	Subicular cortex	CA1 fibers and the entorhinal cortex	Prefrontal cortex Lateral hypothalamus Mamillary nuclei Amygdala Entorhinal cortex Nucleus accumbens	Working memory
Amygdala	Corticomedial group Basolateral group	Olfactory bulb Solitary nucleus Temporal cortex Prefrontal cortex Ventromedial hypothalamus Medial hypothalamus Substantia innominata	Midbrain periaqueductal gray Hypothalamus Stria terminalis Bed nucleus of the stria terminalis	Cardiovascular function Variety of emotional responses, e.g. detachment, relief, relaxation Feeding Endocrine functions Micturition Organization of fear response
Septal area	Bed nucleus of stria terminalis Nucleus accumbens	Hippocampal formation Brainstem (monoaminergic system) Lateral hypothalamus	Hippocampal formation Hypothalamus (medial)	Rage and aggression Autonomic functions

Note that the input and output do not correlate here. It is merely the collections of afferent and efferent output of each territory.

3.5 DIENCEPHALON

The diencephalon is comprised of not only the thalamus and the hypothalamus, but also the subthalamus and epithalamus. Each of these will be discussed below in relation to the major nuclei, what afferent and efferent information it transmits and/or the major functions of each territory.

3.5.1 Thalamus

The thalamus is divided up according to its nuclei – anterior, lateral, medial, midline and intralaminar. Similarly, the thalamus can be classified according to the functional territories – specific, non-specific and association nuclei. The table below will detail the input, output and related functions of the thalamic nuclei in summary form (Table 3.4).

Table 3.4. The Specific Nuclei of the Thalamus in Terms of Their Input, Output and Functions

Thalamus			
Specific Thalamic Nuclei	**Input**	**Output**	**Functions**
Somatosensory Relay Nuclei			
Ventral posterolateral nucleus	Spinothalamic tract and medial lemniscus	Postcentral gyrus (somesthetic cortex)	Touch, pressure and joint movement
Ventral posteromedial nucleus	Trigeminothalamic tract	Postcentral gyrus (somesthetic cortex)	Sensory information from the face Taste
Medial geniculate nucleus	Inferior colliculus	Brodmann's area 41	Hearing
Lateral geniculate nucleus	Retina	Primary visual cortex (Brodmann's area 17)	Vision
Motor Functions			
Ventrolateral nucleus	Dentate nucleus of cerebellum, substantia nigra and medial pallidus	Ventroanterior nucleus, Primary motor cortex (Brodmann's area 4) and premotor cortex (area 6)	Motor function
Ventral anterior nucleus	Substantia nigra and medial pallidus	Premotor and supplementary areas (Brodmann area 6)	Motor function
Limbic System			
Anterior nucleus	Mamillary bodies. Subicular cortex	Cingulate cortex	Memory
Mediodoral thalamic nuclei*	Prefrontal cortex, amygdala, septal area, prepyriform area and anterior cingulate gyrus	Prefrontal cortex and the anterior lateral hypothalamus	Modulation of the hypothalamus Affective components of pain
Integrative Functions			
Lateral posterior thalamic nucleus	Parietal lobe	Superior parietal lobule (Brodmann areas 5 and 7)	Sensorimotor integration
Pulvinar nuclei	Superior colliculus and visual cortex Temporal neocortex	Parietal lobule and temporal lobe	Integration of visual and auditory information Sensory discrimination Cognition
Non-specific			
Centromedian nucleus	Nociceptive fibers from spinal cord Medial pallidus Reticular formation Cerebral cortex	Putamen	Arousal and attention Modification of sensory information
Reticular nucleus	Cortex and dorsal thalamic nuclei	Dorsal thalamic nuclei	Arousal (i.e. sleep/ wakening) Regulation of flow to and from the cerebral cortex
*Note that the mediodorsal thalamus is discussed in a little more detail in the previous chapter.			

3.5.2 Hypothalamus
The hypothalamus is essential for the regulation of autonomic and endocrine regulation within the body. It is essential for a variety of functions including feeding and drinking and water balance in the body. It regulates autonomic control in the body, but has also been implicated in the flight reaction, and with rage and aggression. The hypothalamus is comprised of the ventromedial, paraventricular, suprachiasmatic, supraoptic, arcuate, dorsomedial and tuberal nuclei, as well as the preoptic region and mammillary bodies. A further detailed breakdown of the hypothalamic nuclei and their related functions, and position within the hypothalamus is given below (Table 3.5).

Table 3.5. The Functions of Each of the Regions of the Hypothalamic Nuclei

Hypothalamus	
Region and Related Nuclei	Functions
Anterior Medial	
Supraoptic nucleus	Release of antidiuretic hormone (vasopressin)
Medial preoptic nucleus	Releases gonadotrophin releasing hormone (GnRH)
Anterior hypothalamic nucleus	Sweating, temperature regulation
Suprachiasmatic nucleus	Maintenance of the circadian rhythm
Paraventricular nucleus	Release of oxytocin, corticotrophin releasing hormone, thyrotrophin releasing hormone and somatostatin
Anterior Lateral	
Lateral nucleus	Hunger and thirst
Lateral preoptic nucleus	Maintenance of body temperature
Posterior Lateral	
Lateral nucleus	Hunger and consciousness
Posterior Medial	
Posterior nucleus	Shivering Release of ADH Raising of blood pressure Mydriasis
Mamillary nuclei	Recollection memory

Table 3.6. The Afferent and Efferent Input to and from the Subthalamus by Region, and also Related Functions of These Territories

Subthalamus			
Structure	Afferent	Efferent	Functions
Subthalamic nucleus	Globus pallidus Motor cortex	Globus pallidus	Regulation of motor function in communication with the basal ganglia
Zona incerta	Frontal, parietal and occipital lobes Intralaminar nuclei of the thalamus Basal ganglia (substantia nigra) Raphe nuclei Periaqueductal gray matter Superior colliculus Spinal cord	Cerebral cortex Intralaminar nuclei of the thalamus Anterior and lateral hypothalamus, preoptic area and paraventricualr area Diagonal band of broca Basal ganglia (substantia nigra) Spinal cord	Poorly under-stood but poten-tially linked to integration of the motor and limbic system (visceral pain)

3.5.3 Subthalamus

The subthalamus is located at the posterior one-third of the diencepha-lon and is closely associated with the basal ganglia.

3.5.4 Epithalamus

The epithalamus comprises the thalamus, hypothalamus and also the pituitary gland. It is located at the dorsal aspect of the diencephalon. It also includes the habenula, pineal gland and also the stria medullaris. The thalamus and hypothalamus have been previously described. The tables below highlight these various structures, their components and also the functions of each of the areas (Table 3.6).

3.5.4.1 Pituitary Gland

The pituitary gland is located in the *sella turcica* of the sphenoid bone. It is comprised primarily of an anterior and posterior part. The posterior pituitary part is connected to the hypothalamus by the *median eminence* via the tube-like *pituitary stalk*. The anterior pituitary portion is regulated from the hypothalamus via the *paraventricular nucleus of the hypothalamus*. The posterior pituitary is controlled by the neuroendocrine (*magnocellular* neurosecrteory) cells from the *paraventricular* and *supraoptic nucleus*. The table below highlights the functions of each of the regions of the pituitary gland, the target and the role of the hormones it produces (Table 3.7).

Table 3.7. The Roles of the Anterior and Posterior Pituitary Gland Relating the Hormonal Production to its Target Cells and Related Functions of Each of These

Pituitary Gland		
Region	Target	Function
Anterior		
Adrenocorticotrophic hormone (ACTH)	Adrenal cortex (zona fasciculata)	Increase in the production and release of corticosteroids
Luteinizing hormone (LH)	Supports theca cells	Development of the corpus luteum and stimulation of ovulation
	Leydig cells	Production of testosterone
Follicle stimulating hormone (FSH)	Granulosa cells	Initiation of growth of the follicle
	Sertoli cells	Spermatogenesis
Human growth hormone (HGH)	Growth hormone receptors throughout the body	Cell reproduction, growth and regeneration Increases free fatty acids and glucose
Thyroid stimulating hormone (TSH)	TSH receptors in follicular cells of the thyroid gland	Production of thyroid hormones (T3 and T4) to increase metabolism
Posterior		
Oxytocin	Oxytocin receptor (found in endometrium, myometrium and throughout the central nervous system)	Reproduction
Antidiuretic hormone (ADH)	Distal convoluted tubule and collecting duct	Vasoconstriction Water retention

3.5.4.2 Habenula

Part of the epithalamus is the habenula. It is comprised of two nuclei – the lateral and medial. It is related to the pineal gland, which produces *melatonin* responsible for the circadian rhythm. The habenula has been thought of as evolving in close relation to the pineal gland (Guglielmotti and Cristino, 2006) (Table 3.8).

Table 3.8. The Afferent and Efferent Input and Output of the Habenula and the Associated Functions

Habenula			
Region	Input	Output	Functions
Lateral habenula	Lateral hypothalamus Ventral pallidum	Tegmental nucleus Substantia nigra (pars compacta) Median and dorsal raphe nuclei	Decision making and reward response Reinforcement learning Mood Sleep
Medial habenula	Locus ceruleus Septal nuclei Ventral tegmentum Diagonal band of Broca Preoptic area	Pineal gland Interpeduncular nucleus	Mood Sleep

3.5.4.3 Pineal Gland

The pineal gland is located at the rear of the third ventricle and is an endocrine gland. It receives its input from the superior cervical ganglion of the sympathetic nervous system. It produces melatonin in response to darkness. This lack of light is detected by the retinal cells which project to the *suprachiasmatic nucleus of the hypothalamus*. These fibers then project to the *paraventricular nucleus* which then relays information back to the spinal cord, then onward to the superior cervical ganglia, and then back to the pineal gland.

3.5.4.4 Stria Medullaris

The stria medullaris receives afferent information from the anterior thalamic and septal nuclei as well as the lateral preopticohypothalamic area. It then communicates with the habenula.

REFERENCES

Brodmann, K., 1909. Vergleichende Lokalisationslehere der Grosshirnrinde. Johann Ambrosius Barth, Leipzig.

De la Riva, P., Smith, K., Weintraub, D., 2014. Course of psychiatric symptoms and global cognition in early Parkinson disease. Neurology Published online before print August 15, 2014, doi: 10.1212/WNL.0000000000000801.

Garey, L.J., 2006. Localisation in the cerebral cortex. Springer, New York, http://www.appliedneuroscience.com/Brodmann.pdf (accessed 02.08.2014).

Guglielmotti, V., Cristino, L, 2006. The interplay between the pineal complex and the habenular nuclei in lower vertebrates in the context of the evolution of cerebral asymmetry. Brain Res. Bull. 69, 475–488.

CHAPTER 4

Midbrain (Mesencephalon)

4.1 REVISION POINT

The following is a summary of the main regions of the brain, detailing what comprises the hindbrain (Table 4.1).

(1) *Telencephalon* (cerebral hemispheres) + *Diencephalon* (thalamus and hypothalamus) = FOREBRAIN
(2) *Mesencephalon* = MIDBRAIN
(3) *Metencephalon* (pons, cerebellum and the trigeminal, abducent, facial and vestibulocochlear nerves) + *Myelencephalon* (medulla oblongata, plus the glossopharyngeal, vagus, accessory and hypoglossal nerve nuclei) = HINDBRAIN

The midbrain connects the forebrain and the hindbrain. It is found in the tentorial notch of the dura mater. It is comprised of a *ventral part*, the *cerebral peduncles* and the *tectum*.

The cerebral peduncles are two large bundles which converge from the cerebral hemispheres, and is continuous with the internal capsule. The anterior part of the cerebral peduncles is called the *crus cerebri* and the posterior portion is referred to as the *tegmentum*. The superior part of each peduncle is crossed by the optic tract. The left and right optic tracts emerge from the optic chiasma, which is formed by the junction of the two optic nerves. The depression posterior to the chiasma and bounded by the optic tracts and the cerebral peduncles is called the *interpeduncular fossa*.

The interpeduncular fossa contains the tuber cinereum and the infundibular stem of the hypophysis, the mammillary bodies and the posterior perforated substance. The oculomotor nerve emerges at the superior border of the pons and the medial border of the corresponding cerebral peduncles.

Essential Clinical Anatomy of the Nervous System. http://dx.doi.org/10.1016/B978-0-12-802030-2.00004-2

Table 4.1. Main Components of the Midbrain In Terms of the Afferent and Efferent Connections of Each of the Territories and An Overview of Their Function

Midbrain			
	Afferent	Efferent	Functions
Tectum			
(a) Superior colliculus	Retina Cortical visual association areas Frontal eye fields Pain and auditory pathways	Visual association areas via the pulvinar	Integration of visual and motor functions of the eye
(b) Inferior colliculus	Brainstem nuclei Auditory cortex Cochlear nucleus Medial geniculate body	Medial geniculate body	Convergence of auditory systems Vestibulo-ocular reflex
Cerebral peduncles			
(a) Substantia nigra			
Pars compacta	Inhibitory input from the pars reticulata	Excitatory (D2 receptors) and inhibitory (D1 receptors) input to the neostriatum via dopamine	Prediction of reward with behavior Motor control
Pars reticulata	Inhibitory influence (via GABA) from the neostriatum	Inhibitory output (via GABA) to the superior colliculus and the ventrolateral and ventral anterior thalamic nuclei	Control of eye movements and orientation
(b) Midbrain tegmentum	Multisynaptic connections		Reflex pathways Homeostasis
(c) Crus cerebri	Corticopontine fibers		Cerebellar function

The tectum is comprised of four hillocks called the superior and inferior colliculi. The superior one is concerned with visual function, and the inferior colliculus is responsible for the transmission of auditory information. The pineal gland is attached to the forebrain above the superior colliculi. The midbrain is also traversed by the cerebral aqueduct linking the third ventricle above to the fourth ventricle below.

CHAPTER 5

Hindbrain (Rhombencephalon)

5.1 REVISION POINT

As previously discussed, the brain is subdivided as follows.

(1) *Telencephalon* (cerebral hemispheres) + *Diencephalon* (thalamus and hypothalamus) = FOREBRAIN
(2) *Mesencephalon* = MIDBRAIN
(3) *Metencephalon* (pons, cerebellum and the trigeminal, abducent, facial and vestibulocochlear nerves) + *Myelencephalon* (medulla oblongata plus the glossopharyngeal, vagus, accessory and hypoglossal nerve nuclei) = HINDBRAIN

The hindbrain, comprised of the metencephalon and the myelencephalon, will now be discussed in more detail.

5.2 METENCEPHALON

5.2.1 Pons

The pons is found between the medulla and the midbrain. It lies anterior to the cerebellum and superficially bridges the two cerebral hemispheres. From viewing it anteriorly, transverse fibers at the front of it form the *middle cerebellar peduncle* on each side and enter the cerebellum. The fibers within this connect one cerebral hemisphere with the contralateral cerebellar hemisphere.

The anterior portion of the pons lies on the basilar portion of the occipital bone, and is grooved anteriorly where the basilar artery is found. The sixth (abducent), seventh (facial) and eighth (vestibulocochlear) cranial nerves emerge from between the pons and medulla. Approximately half way up in the pons, the trigeminal nerve (motor and sensory roots) emerges from the lateral aspect of the pons.

Essential Clinical Anatomy of the Nervous System. http://dx.doi.org/10.1016/B978-0-12-802030-2.00005-4

The posterior surface of the pons forms part of the floor of the fourth ventricle, and is laterally bounded by the superior cerebellar peduncles. The fibers within this connect the cerebellum with the midbrain.

5.2.2 Cerebellum

The cerebellum is found on the posterior aspect of the brainstem (in the posterior cranial fossa) and is attached to it by three *cerebellar peduncles*. The superior cerebellar peduncle connects with the midbrain, the middle with the pons and the inferior with the medulla. It comprises a median portion called the *vermis* and two lateral portions called the *cerebellar hemispheres*. The cerebellum is comprised of a cortex of gray matter and is folded to form *folia*, like the cerebral hemispheres sulci. The cerebellum is connected to the cerebral cortex and the spinal cord via tracts. Deep in the cerebellum's white matter are four nuclei – fastigial, globose, emboliform and the dentate nuclei. The cerebellum is essential for learning and executing movement and ensures coordination of muscle contraction force and the extent and duration of this process. In addition, the cerebellum also has a role to play in cognition and general affect, including happiness (Schienle and Scharmüller, 2013).

Within the cerebellar cortex, there are three layers – the molecular, Purkinje and the granule cell layer. These are summarized in terms of their input and output below (Table 5.1).

In addition, the cerebellum is comprised of three peduncles, and their input and output connections are shown below (Table 5.2).

Table 5.1. The Cell Type and the Afferent and Efferent Input to the Cortex of the Cerebellum			
Cortex of Cerebellum			
Layer	Cell Type	Afferent	Efferent
Molecular	Basket	Climbing fibers to the dendritic tree of Purkinje cells Golgi cell dendrites Granule cell dendrites	Purkinje cell layer from the basket cells
Purkinje	Purkinje	Basket cell dendrites	Molecular layer and Golgi cells Central nucleus of cerebellum
Granule cell	Granule	Purkinje cells Golgi cell targeting granule cells Mossy fibers	Dendritic tree of Purkinje cells Axon terminals of Golgi cells

Table 5.2. The Afferent and Efferent Input to the Cerebellar Peduncles

	Cerebellum	
Peduncle	**Afferent**	**Efferent**
Superior	Locus coerulues Trigeminal nucleus (mesencephalic) Superior and inferior colliculi Spinocerebellar tract (ventral)	Contralateral ventral lateral nucleus of the thalamus Red nucleus (ipsilateral)
Middle	Pontocerebellar fibers	N/A
Inferior	Arcuate nucleus Spinocerebellar tract (dorsal) Precerebellar reticular nucleus Vestibular nerve (and its nucleus) Olivocerebellar fibers Cuneocerebellar fibers Trigeminal nuclei (spinal and pontine)	Vestibular nuclei Reticular nuclei

5.3 CLINICAL ASSESSMENT

There are several aspects to examining the integrity of the cerebellum.

(1) Always introduce yourself to the patient (in any clinical examination or history taking) and state your position.

(2) *Assess gait.* Ask the patient to walk from one side of the room (or examining area) to the other. If they normally use an aid to walking, they should be allowed to do so.

(3) *Heel to toe.* The patient should be asked to walk forward by placing one heel in front of the toes then switching to the opposite side and to keep walking in this fashion for a short distance

(4) *Romberg's test.* Further details are also found in Chapter 8.
   (a) Ask the patient to stand up with their feet together, arms by their side and eyes open.
   (b) Then, ask the patient to close their eyes for approximately 20–30 s.
   (c) The patient may exhibit mild swaying which is normal.
   (d) It is possible to repeat the test two times to help assessment. If the patient loses their balance, it is said that they have a positive Romberg's test, or Romberg's sign.

(5) Check for a *resting tremor* by having the patient place their arms and hands out straight.

(6) *Assess muscle tone and power* as discussed in detail in Chapter 8.

(7) *Check for dysdiadochokinesis.* Ask the patient to touch one dorsal surface of the hand with the palmar surface of the opposite hand. The opposite hand should then rotate to the dorsal surface of the opposite hand. This alternating palmar/dorsal surface onto the opposite hand should be repeatedly as rapidly as possible for the patient. Dysdiadochokinesis is the inability to undertake this rapid movement.

(8) *Finger to nose.* The patient should touch their nose then the examiners finger which is held in space. The examiner should move their examining finger and the patient should repeat the movement of touching their nose and the moving examiners finger.

(9) *Heel to shin test.* The patient should be asked to place the heel of one foot at the knee of the opposite leg. Then roll the heel down the front of the shin and back up. This should be repeated several times. Repeat this on the opposite side several times too.

TIP!

The following mnemonic of D.A.N.I.S.H. is a helpful reminder of the key features when there is cerebellar disease.

D – *D*ysdiadochokinesis
A – *A*taxia
N – *N*ystagmus
I – *I*ntention tremor
S – *S*lurred speech
H – *H*ypotonia

5.4 PATHOLOGIES

There are a variety of causes of cerebellar damage. These are listed in the following table, split into the various categories they can be classified into (Table 5.3).

5.4.1 Related Cranial Nerves

The trigeminal, abducent, facial and vestibulocochlear nerves are found within the metencephalon. They are discussed in more detail in Chapter 1.

5.4.2 Myelencephalon
5.4.2.1 Medulla Oblongata

The superior aspect of the spinal cord enters through the foramen magnum to become the medulla oblongata. It rests on the basilar

Table 5.3. A Summary of the Various Causes of Cerebellar Disease

	Cerebellar disease
Neurological	Multiple sclerosis
Metabolic	Wilson's disease Hypothyroidism Mitochondrial disorders
Congenital	Cerebral palsy Cerebellar hypoplasia Dandy–Walker syndrome
Genetic	Friedrich's ataxia Ataxia telangiectasia
Drugs	Phenytoin Barbiturates Overdose of temazepam
Infections	Meningo-encephalitis Intracranial abscess
Nutritional	Thiamine deficiency Deficiency of vitamin E Gluten sensitivity
Trauma	Various

portion of the occipital bone, separated from it by the left and right vertebral arteries. On the posterior aspect of the medulla oblongata, the cerebellum covers it.

The inferior area of the medulla also contains the continuation of the central canal of the spinal cord, and at its superior aspect, becomes the fourth ventricle.

The medulla has an *anterior median fissure,* and the lower portion is interrupted by the decussation of the pyramids, where the descending pyramidal fibers cross the median plane. The portion of the medulla adjacent to the superior aspect of the anterior median fissure on either side is the *pyramid.* The pyramids contain the fibers of the pyramidal (*corticospinal*) tract. On either side of the pyramids, an elevation is present called the olive, which is bounded by the anterolateral sulcus medially and the posterolateral sulcus on its lateral aspect.

The hypoglossal nerve arises from the medulla between the olive and the pyramid. The glossopharyngeal, vagus and accessory nerves arise posterolateral to the olive.

On the dorsal region of the medulla, there is a posterior median sulcus. On either side, two tracts from the spinal cord terminate in eminences called the gracile and cuneate tubercles. These arise from the gracile and cuneate fasciculi, which are described in more detail in Chapter 8, Section 8.1. Superiorly, the lower part of the fourth ventricle is bounded laterally by the inferior cerebellar peduncle which connects the medulla and the spinal cord with the cerebellum.

The medulla oblongata is responsible for a variety of autonomic related functions including respiration and regulation of the cardiac center. It is also responsible for deglutition, vomiting, sneezing and coughing, all of which are reflexes. The medulla is also the site of afferent and efferent pathways which pass through it. As such, the historical name for hat was bulb, or bulbar, and referred to the tracts and nerves which were connected here and passed through it. Bulb tends not to be used nowadays, but the term bulbar palsy is still kept. Bulbar palsy refers to pathologies involving the nuclei and pathways associated with the glossopharyngeal, vagus, accessory and hypoglossal nerves which are found within the medulla oblongata. The details of these cranial nerves in terms of their anatomy and functions can be found in Chapter 1, and bulbar palsy will be discussed later in this section.

The medulla is the site of ascending and descending tracts, and these will be briefly mentioned in the table below (Table 5.4), including their broad functions, but will be discussed in more detail in subsequent chapters, as indicated.

5.4.2.2 Related Cranial Nerves
The glossopharyngeal, vagus, accessory and hypoglossal nerves are the related cranial nerves and are covered in more detail in Chapter 1.

5.5 CLINICAL ASSESSMENT

A variety of pathways pass through the medulla oblongata as mentioned previously. These include motor and sensory pathways, and they are discussed in more detail in the relevant chapters mentioned in Table 8.4.

Clinical assessment of the cranial nerves found within the medulla, i.e. the glossopharyngeal, vagus, accessory and hypoglossal nerves, is discussed in Chapter 1.

Table 5.4. The Ascending and Descending Tract Functions That Pass Through the Medulla Oblongata

Medulla Oblongata Tracts		
Tract	Function	Chapter for More Discussion
Ascending		
Medial lemniscus system	Discriminative and fine touch, proprioception and vibration	Chapter 8, Section 8.2
	This is the site of crossover from one side of the body to the other	
Spinocerebellar tract	Proprioception from limbs	Chapter 8, Section 8.3
Spinothalamic tract	Pain Temperature Light touch Firm pressure	Chapter 8, Section 8.4.1
Spinoreticular tract	Perception of pain Sleep – arousal Somatic motor function Visceral activity	Chapter 8, Section 8.4.3
Spinotectal (spinomesencephalic) tract	Modulation of pain	Chapter 8, Section 8.4.4
Descending		
Corticospinal tract	Motor function	Chapter 9, Section 9.1.1
Rubrospinal tract	Motor function	Chapter 9, Section 9.8.5
Tectospinal tract	Small in humans Control of head and eye movement	Chapter 9, Section 9.8.6
Central tegmental tract	Motor function and gait	
Nuclei of trigeminal, vestibulocochlear, vagus, accessory and hypoglossal nerves	Various functions	Chapter 1
Also cross-references to the relevant chapters for more detail.		

5.6 PATHOLOGIES

5.6.1 Bulbar Palsy

A bulbar palsy refers to disease affecting the glossopharyngeal, vagus, accessory and hypoglossal nerves and is due to lower motor neuron pathology. Typically, patients with a bulbar palsy present with signs and symptoms of the cranial nerves affected as mentioned. The patient will have dysphagia, dysarthria, flaccid pareses, atrophy and fasciculation of muscles supplied by those cranial nerves and fibrilliation of the tongue (Kühnlein et al., 2008). In addition, the patient would have weakness of the palate and reduced or absent gag reflex, dribbling of saliva and a nasal speech.

The causes of a bulbar palsy vary and can include vascular (infarction of the medulla), degenerative disease (amyotrophic lateral sclerosis (Kühnlein et al., 2008), syringobulbia), malignancy (of the brainstem), inflammation (e.g. poliomyelitis, Guillain-Barré) or genetic disease (Kennedy's disease (NIH, 2014)).

REFERENCES

Kühnlein, P., Gdynia, H.-J., Soerfeld, A.-D., Linder-Pfleghar, B., Ludolph, A.C., Prosiegel, M., Riecker, A., 2008. Diagnosis and treatment of bulbar symptoms in amyotrophic lateral sclerosis. Nat. Clin. Pract. Neurol. 4, 366–374.

National Institute of Health, National Institute of Neurological Disorders and Stroke. http://www.ninds.nih.gov/disorders/kennedys/kennedys.htm (accessed 14.08.2014).

Schienle, A., Scharmüller, W., 2013. Cerebellar activity and connectivity during the experience of disgust and happiness. Neuroscience 246, 375–381.

CHAPTER 6

Blood Supply of the Brain and Clinical Issues

6.1 GENERAL INTRODUCTION

The blood supply of the brain arises from two main arteries – the *internal carotid* and the *vertebral arteries*. The left and right internal carotid and the left and right vertebral arteries merge on the undersurface of the brain. It is referred to as the *Circle of Willis* (or circulus arteriosus cerebri (Willisii) (Symonds, 1955; Eastcott, 1994)). It is named after Thomas Willis who described it after extensive studies of the brain and nerves.

The arterial supply is also classified as anterior and posterior. The anterior cerebral supply arises from the internal carotid arteries and the posterior cerebral supply arises from the vertebral arteries. The *posterior cerebral communicating arteries* (or *posterior communicating arteries*) unite these two supplies. Not everyone has a complete Circle of Willis, and approximately only a quarter of the population have a fully intact Circle of Willis (Creasy, 2011; Hines and Marschall, 2012), with collateral pathways existing in those without one. The *anterior communicating arteries* unite the two anterior cerebral arteries.

6.2 INTERNAL CAROTID ARTERY

The common carotid artery arises from different points on the left and right sides of the body. On the left hand side, the common carotid artery arises direct from the aorta and ascends into the neck alongside the phrenic and vagus nerves. As the common carotid artery ascends into the neck, it bifurcates at approximately the level of the third or fourth cervical vertebra into the external and internal carotid arteries. The other way to look at it clinically is that the bifurcation occurs at the level of the upper border of the thyroid cartilage. This part of the internal carotid artery can also be referred to as its *cervical segment*. At the point of bifurcation, and start of the internal carotid artery, it is at this point where the carotid sinus is found. Baroreceptors are found in

Essential Clinical Anatomy of the Nervous System. http://dx.doi.org/10.1016/B978-0-12-802030-2.00006-6

the internal carotid artery at its point of origin, which are pressure sensitive receptors essential for maintaining blood pressure. As the internal carotid artery ascends the neck, on its way to the internal cranial cavity, it does not give off any branches.

The internal carotid artery then passes to the base of the skull. It passes into the cranial cavity via the carotid canal which lies in the *petrous part of the temporal bone*. This portion is referred to as its *petrous segment* (supplying the pterygoid canal and the tympanic cavity via the vidian and caroticotympanic arteries, respectively). The internal carotid artery lies close to the *foramen lacerum*, referred to as its *lacerum segment*.

During its winding course it then passes into its *cavernous segment* inside the *cavernous sinus*, and its curved portion at that point is called the carotid siphon. This would supply the *cavernous sinus* and the *pituitary gland*. It passes from the medial side of the *anterior clinoid process* (*clinoid segment*) to enter the cranial cavity via the *subarachnoid space*. From here it passes close to the optic nerve (*ophthalmic segment* – dealt with later) and then on to the medial aspect of the lateral cerebral sulcus, or lateral cerebral fissure (*communicating segment* – terminating as the posterior communicating and anterior choroidal arteries, dealt with later).

At the medial side of the lateral cerebral sulcus (or Sylvian fissure), two major arteries arise from the internal carotid artery namely the anterior and middle cerebral arteries. The following table (Table 6.1.) show the major branches of the internal carotid artery, their sub-branches and what they supply.

The other way to look at it is that the internal carotid artery has an ophthalmic branch, posterior communicating, choroidal, anterior cerebral and middle cerebral arteries. First, the branches of the ophthalmic artery (from the ophthalmic segment) are given below.

Other branches of the internal carotid artery (and what they supply) include the *caroticotympanic* (tympanic membrane), *pterygoid* (inconstant; pterygoid canal), *cavernous* (trigeminal ganglion), *hypophysial* (pituitary gland) and *meningeal* (dura mater of anterior cranial fossa) branches.

Table 6.1. Branches of the Ophthalmic Artery and the Anatomical Territories Supplied By Them

Major Branch of Ophthalmic Segment	Sub-branches	Territories Supplied
Ophthalmic artery	Central artery of the retina	Optic nerve Inner layers of the retina from its four branches (superior and inferior nasal and temporal branches)
	Lacrimal artery	Lacrimal gland Eyelids and conjunctiva Extraocular muscles (variable)
	Muscular branches	Extraocular muscles of the eye
	Ciliary arteries	Choroid coat Ciliary processes
	Supraorbital arteries	Skin, muscles and pericranium of the forehead Levator palpebrae superioris Superior rectus Trochlea Frontal bone Frontal sinus (inconstant)
	Posterior ethmoidal artery	Posterior ethmoidal sinus Dura mater Nasal cavity
	Anterior ethmoidal artery	Middle and anterior ethmoidal sinuses Frontal sinus Dura mater Nasal cavity
	Meningeal branch	Bone of middle cranial fossa
	Medial palpebral arteries	Eyelids Nasolacrimal duct
	Supratrochlear artery	Skin, muscles and pericranium of the forehead
	Dorsal nasal artery	Lacrimal sac Dorsum of the nose (with anastomoses with the facial artery)

6.2.1 Posterior Communicating Artery

This small artery originates from the terminal bifurcation of the internal carotid artery. It passes posteriorly to join with the posterior cerebral artery hence forming the Circle of Willis.

6.2.2 Anterior Choroidal Artery

The anterior choroidal artery arises from the distal part of the internal carotid artery. From its point of origin, the anterior choroidal artery passes along the optic tract and the temporal lobe, at the choroid fissure at its medial edge. The following table (Table 6.2) highlights the extensive distribution that this artery supplies.

Table 6.2. The Areas Supplied By the Anterior Choroidal Artery

Territories supplied by the anterior choroidal artery
Hippocampus
Uncus
Optic tract
Globus pallidus
Internal capsule (ventral portion)
Choroid plexus of the lateral ventricle

The termination of the internal carotid artery is as the *anterior* and *middle cerebral arteries*. First the anterior cerebral artery, its sub-branches and what they supply will be dealt with. The middle cerebral artery, its branches and the territories that it supplies will follow this.

6.2.3 Anterior Cerebral Artery

The anterior cerebral artery is the smaller of the two terminal branches. It commences at the medial aspect of the Sylvian fissure passing anterior and medial to the optic nerve and is closely related to the anterior cerebral artery of the opposite side, and is joined to it by the small *anterior communicating artery*. The anterior cerebral artery has cortical and central branches which will be dealt with below. The *recurrent artery of Heubner* also arises from the anterior cerebral artery. This branch supplies the internal capsules anterior limb, caudate nucleus, and putamen (Table 6.3).

Table 6.3. The Territories Supplied by the Central and Cortical Branches of the Anterior Cerebral Artery

Anterior Cerebral Artery	Sub-branches	Territories Supplied
Central artery	Many small arterial branches	Corpus callosum (rostrum) Septum pellucidum Caudate nucleus (head) Anterior putamen (of lentiform, or lenticular nucleus)
Cortical branches	Orbital branches	Frontal lobe (orbital surface): Olfactory lobe Medial orbital gyrus Straight rectus
	Frontal branches	Medial frontal gyrus Cingulate gyrus Paracentral lobule Superior frontal gyrus Middle frontal gyrus Precentral gyrus
	Parietal branches	Precuneus (superior parietal lobule) and nearby lateral surface

Table 6.4. The Territories Supplied By the Central and Cortical Branches of the Middle Cerebral Artery

Middle Cerebral Artery	Sub-branches	Territories Supplied
Central artery	Medial (lenticulo)striate arteries*	Internal capsule (anterior limb (inferior aspect) to genu) Lentiform nucleus Caudate nucleus Anterior commissure (midline fibers) Anterior hypothalamus Medial forebrain bundle Olfactory bulbs Optic chiasm
	Lateral (lenticulo)striate arteries*	Internal capsule (anterior and posterior limbs (superior aspect) Lentiform nucleus Caudate nucleus (superior and lateral two-thirds of head and body) Putamen Globus pallidus (lateral) Anterior commissure (lateral one-third) Anterior perforated substance
Cortical branches	Frontal branches	Frontal gryrus (middle and inferior) Precentral gyrus
	Orbital branches	Inferior frontal gryrus Frontal lobe (orbital surface)
	Parietal branches	Parietal lobe
	Temporal branches	Temporal lobe (lateral surface)

*Jośe (2009). Note the medial (lenticulostriate) arteries have also been referred to as the recurrent artery of Heubner (named after the German pediatrician who described it – Otto Heubner).

6.2.4 Middle Cerebral Artery

The middle cerebral artery is the larger of the two terminations of the internal carotid artery. It passes in the lateral aspect of the Sylvian fissure. It passes posterior and superiorly on the surface of the insula providing branches to it, and also to the lateral aspect of the cerebral hemispheres on each side it arises.

The middle cerebral artery has two main branches – cortical and central. Their branches and distribution can be seen in the following table (Table 6.4).

6.3 VERTEBRAL ARTERY

The vertebral artery is the other origin of the arterial supply of the brain. It arises from the subclavian artery. It passes into the transverse foramen (foramen transversarium) of the upper six cervical vertebrae.

Table 6.5. The Territories Supplied By the Vertebral and Basilar Arteries Based on Their Branches

	Sub-branches	Territories Supplied
Vertebral artery	Spinal branches/arteries	Spinal cord and its membranes Bodies and periosteum of vertebrae
	Muscular branches	Deep neck muscles
	Cranial branches	Meningeal branches – falx cerebelli Anterior and posterior spinal artery – spinal cord Posterior inferior cerebellar artery – cerebellum (posterior), choroid plexus of fourth ventricle, central nuclei of the cerebellum, inferior vermis and posterolateral aspect of the medulla
Basilar artery	Labyrinthine artery	Internal ear
	Pontine branches	Pons
	Anterior inferior cerebellar artery	Anterolateral inferior surface of cerebellum Inferolateral aspect of the pons
	Superior cerebellar artery	Cortex, white matter and central nuclei of the posterior (dorsal) aspect of the cerebellum Pons Pineal body Third ventricle Superior medullary velum
	Posterior cerebral artery	See below

At the base of the skull, the vertebral artery is intimately associated with the atlas (C1 vertebra) and passes through the posterior atlanto-occipital membrane. At the foramen magnum it enters the subarachnoid space by passing through the dura and arachnoid mater. It courses over the medulla and the left and right vertebral arteries then pass medial to unite to form the basilar artery on the pons. The termination of the basilar artery is as the posterior cerebral arteries.

There are several branches of the vertebral and basilar arteries which are important to note. These are discussed above detailing their sub-branches and what territories they supply (Table 6.5).

6.4 POSTERIOR CEREBRAL ARTERY

The posterior cerebral artery passes lateral after coming from the basilar artery. It winds round the cerebral peduncle passing toward the tentorial aspect of the cerebrum. It has numerous small branches as it goes on to supply the occipital and temporal lobes. The main branches of the posterior cerebral artery are divided broadly into the central and cortical divisions, as described below (Table 6.6).

Table 6.6. The Territories Supplied By the Central and Cortical Branches of the Posterior Cerebral Arteries

	Sub-branches	Territories Supplied
Central branches	Posteromedial branches	Globus pallidus Third ventricle (lateral wall) Thalamus (anterior)
	Posterior choroidal artery	Lateral geniculate body Lateral ventricle (posteroinferior horn) Posterior thalamus Choroid plexus of the lateral and third ventricles
	Posterolateral central branches	Medial geniculate body Posterior thalamus Cerebral peduncle Colliculi
Cortical branches	Temporal branches	Occipitotemporal gyri (lateral, medial and parahippocampal) Uncus
	Occipital branches	Posterolateral aspect of the occipital lobe Lingual gyrus Cuneus
	Parieto-occipital	Precuneus Cuneus

6.5 VENOUS DRAINAGE OF THE BRAIN

The veins which drain the brain do not contain valves and have extremely thin walls. They open into the venous sinuses of the dura by piercing through the arachnoid and dura (meningeal layer).

Blood from the brain drains into structures called venous sinuses. These are located between the layers of the dura (meningeal and endosteal). These veins drain ultimately into the internal jugular veins on the way back returning deoxygenated blood to the heart. They are extremely thin walled and are lined by endothelium which is continuous with the veins of the brain and the internal jugular vein. The dural venous sinuses are rather complex running a tortuous course draining the brain. Each one will now be dealt with in turn.

(1) Superior sagittal sinus

This venous sinus commences at the crista galli passing posteriorly and superiorly in the superior aspect of the falx cerebri separating the two cerebral hemispheres. As it reaches the posterior aspect of the cranial cavity at the internal occipital protuberance, it either enters a common venous drainage area with the straight sinus. Sometimes, the superior sagittal sinus will divide into a left and right division, and unite with the transverse sinus of either the left or right sides, respectively. The superior cerebral vein, draining the upper, middle and outer aspects of the brain, will drain into the superior sagittal sinus.

(2) Confluence of the sinuses

This territory is found at the internal occipital protuberance. It is the point where the straight and superior sagittal sinuses end and the transverse sinuses begin. There is, however, wide variation at this site (Park et al., 2008).

(3) Inferior sagittal sinus

The inferior sagittal sinus lies in the lower edge of the falx cerebri. It receives blood from the great cerebral vein (of Galen). The great cerebral vein drains the deep cerebral veins. The inferior sagittal sinus then empties into the straight sinus.

(4) Straight sinus

The straight sinus is a single vessel and is found at the folds of two dural folds – the falx cerebri and the tentorium cerebelli. The straight sinus is formed by the great cerebral vein (of Galen), superior cerebellar veins, inferior sagittal sinus and posterior cerebral vein. It then joins with the confluence of the sinuses.

(5) Transverse sinus

Generally, one of the transverse sinuses predominates, and it is the right side which is found to do so more. This means that generally, there is a greater capacity of the right-sided transverse sinus for capacity of higher blood flow. The transverse sinus passes outwards and anteriorly and is then named the sigmoid sinus at the point it turns at the petrous temporal bone. Emissary veins join the transverse sinuses with the venous plexus of the suboccipital region.

(6) Sigmoid sinus

The sigmoid sinus is a direct continuation of the transverse sinus. It passes inferiorly and medially and courses over the mastoid portion of the temporal bone. The sigmoid sinus terminates at the level of the jugular foramen, and at that point then drains into the internal jugular vein.

(7) Cavernous sinus

The cavernous sinus is a collection of venous channels. It drains the superior ophthalmic and superficial middle cerebral veins and the sphenoparietal sinus. It communicates with the internal jugular vein and transverse sinus via the petrosal sinuses (superior and inferior). The pterygoid plexus provides a means by where the cavernous sinus communicates with the facial vein via emissary veins. Intracavernous sinuses allow for communication between the left and right-sided cavernous sinuses.

It is located between the superior orbital fissure anteriorly, the petrous temporal bone posteriorly and the sphenoid bone laterally. The cavernous sinus opens anteriorly at the level of the superior orbital fissure. As well as being a dural venous sinus, the cavernous sinus contains the *internal carotid artery*, *abducent nerve* and *sympathetic plexus* medially. On the lateral aspect of the cavernous sinus, the oculomotor, trochlear, ophthalmic and maxillary divisions of the trigeminal nerve. However, although these nerves pass through the cavernous sinus, they are separated from the blood by the endothelium.

(8) Petrosal sinuses

There are two petrosal sinuses – an *inferior* and *superior*. The *inferior petrosal sinus* is found between the occipital bone and the petrous temporal bone, in a groove especially for it to sit on. As it passes down to its point of termination in the internal jugular foramen at the jugular foramen, it also receives labyrinthine vessels.

The *superior petrosal sinus* drains the cavernous sinus, passing into the cavernous sinus. It passes in a posterolateral direction in the *tentorium cerebelli*. It drains into the transverse sinus at the point that it then changes into the sigmoid sinus.

(9) Basilar plexus

The basilar venous plexus interconnects the two inferior petrosal sinuses at the clivus. It then joins the superior petrosal sinuses and the cavernous sinus.

(10) Sphenoparietal sinus

The sphenoparietal sinuses receive blood from the dura mater, superficial middle cerebral and temporal diploic veins. It drains into the anterior portion of the cavernous sinus.

(11) Middle meningeal veins
      The middle meningeal veins are found on the inner aspect of the
      parietal bone, and frequently result in impressions on the bone.
      These are generally more laterally than the accompanying middle
      meningeal artery. The middle meningeal veins join superiorly
      with the superior sagittal sinus. In addition, the inferior cerebral
      and superficial middle cerebral veins and diploic veins also drain
      into this vessel. They can pass into the pterygoid venous plexus,
      though their drainage is variable.

(12) Emissary veins
      The emissary veins are valveless vessels which connect the
      superficial veins of the scalp with deeper veins, e.g. diploic veins
      of the skull bones. Emissary veins can be found in a variety of
      locations and can include the mastoid process, parietal bone
      (passing through the large parietal foramen), around the internal
      carotid artery and the cavernous sinus, occipital protuberance,
      foramen lacerum, and potentially around the ophthalmic veins. If a
      patient has an infection in the loose connective tissue of the scalp,
      it may spread via the emissary (and diploic) veins into the skull,
      and also forward onto the face. Spread of infection from the scalp
      may cause septic thrombophlebitis of the emissary veins (Roos and
      Tunkel, 2010). This infection may then spread intracranially and
      rarely, this septic thrombosis of emissary veins can result in venous
      necrosis between the dura and the skull resulting in an epidural
      hemorrhage (Rajput and Rozdilsky, 1971; Moonis et al., 2002).

(13) Diploic veins
      These veins are found within the diploe of the skull. These structures
      are valveless and are dilated at intervals along their distribution, and
      communicate with nearby diploic veins. The diploic veins develop
      after birth and communicate with the scalp veins, meningeal veins as
      well as the dural venous sinuses in close proximity.

6.6 CLINICAL ASSESSMENT

It is absolutely essential that local and national guidelines be followed
for your area/hospital/surgery/country, especially if there is an acute situation, e.g. stroke, transient ischemic attack (TIA) or other emergency
situation (see Section 6.7 later) (NIH, 2014).

In the UK, the National Institute for Health and Care Excellence (2014a) provide appropriate guidance dependent on the possible diagnoses and what procedures need to be followed. It is imperative that, whatever country you (the reader) are in, up-to-date guidelines are adhered to, to ensure the patient is investigated and managed appropriately.

One simple test to assess if a stroke or TIA has occurred is F.A.S.T. The following questions need to be answered quickly (National Health Service, Stroke – Act F.A.S.T.) (NHS, 2013). It asks the following questions:

"F – Face. Has their face fallen on one side? Can they smile?
A – Arms – Can they raise both arms and keep them there?
S – Speech – Is their speech slurred?
T – Time to call 999 (UK emergency services number) if you see any single one of these signs"

This simple test can easily be performed in any setting, clinical or otherwise.

If there is a positive F.A.S.T., and stroke is suspected, rapid diagnosis should be undertaken. Ensure local and national guidelines are adhered to. In the UK, these are in the form of those issued by the National Institute for Health and Care Excellence (2014b), e.g. CG68 Stroke: algorithm 2. They may be different for your country or area of practice, so please consult local and national policy.

If, e.g. the F.A.S.T. was positive, diagnosis should be rapid and can follow a validated tool like the ROSIER scale. This is the Recognition of Stroke in the Emergency Room and uses a seven item score (-2 to +5) and assesses "clinical history (loss of consciousness and convulsive fits) as well as neurological signs (face, arm, or leg weakness, speech disturbance, visual field defect)" (Nor et al., 2005). This can be used to guide if brain imaging is necessary, and if further treatment and admission to a specialist stroke unit is necessary.

If a TIA is suspected and a stroke is ongoing, the above guidance could be used, and if the neurological symptoms have fully resolved, and a TIA is compatible with the history the National Institute for Health and Care Excellence CG68 Stroke: algorithm 1 could be followed for further management (Stroke, 2014).

For patients with a suspected TIA, their likelihood to develop a full stroke should be urgently assessed. This takes into account the patient's age, blood pressure, type of symptoms, if diabetes is present and how long symptoms lasted using the $ABCD^2$ score (Johnston et al., 2006).

If there is a high risk of stroke developing, patients should be commenced on aspirin and seen by a stroke specialist within 24 h. This may require follow-up with an MRI (or CT scan of the brain if MRI is contraindicated) to identify the exact area of the brain involved. If a lower risk of stroke is suspected, the patient should also be commence don aspirin and seen by a stroke specialist within one week (NICE guidelines, CG68, 2008). It may be relevant, if clinically suspected, to undertake a carotid artery ultrasound, to identify those candidates that may be suitable for *carotid endarterectomy*. Whether or not narrowing of the carotid artery is identified, antiplatelet drugs should be given to reduce the incidence of blood clots developing.

6.6.1 Cerebral Angiography

Previously, if imaging were needed of the cerebral vasculature, a cerebral angiogram would be performed. This would involve injecting radio-opaque iodine into either the internal or common carotid artery. It would allow for a clear demonstration of the cerebral vasculature, perhaps when investigating cerebral tumors or detection of aneurysms. It primarily showed the anterior and middle cerebral vessels, but also could potentially demonstrate the posterior circulation too. These images generally were collected approximately 2 s after the injection. Then, after another 2 s or so, the dye would pass to the veins of the brain and a venogram could be obtained. After another 2 s or so, the dye would pass to the venous sinuses and a sonogram could be obtained. However, nowadays, a variety of imaging techniques are employed to accurately identify the cerebral vasculature (including the lenticulostriate arteries), with digital reconstruction from magnetic resonance angiography (Wright et al., 2013; Kang et al., 2005). Indeed, newer techniques have been developed in MR imaging to use enhanced resolution in examining the smallest of blood vessels without the use of contrast medium (Kloppenborg et al., 2011).

6.7 PATHOLOGIES

6.7.1 Cerebrovascular Disease

Cerebrovascular disease is a collection of pathologies which affects the cerebral circulation. Simply put, there are four common pathologies which are classified as cerebrovascular disease (NHS, 2014a):

(1) Stroke – Interruption of the brain's blood supply.
(2) Transient ischemic attack (TIA) – temporary reduction in the supply of blood to the brain.
(3) Subarachnoid hemorrhage – a less common cause for a patient to have a stroke.
(4) Vascular dementia – insufficient blood supply to regions of the brain.

6.7.2 Stroke

According to the Royal College of Physicians (UK) National Clinical Guidelines for Stroke (2012) document, a stroke is defined as "a clinical syndrome, of presumed vascular origin, typified by rapidly developing signs of focal or global disturbance of cerebral functions lasting more than 24 h or leading to death (World Health Organization, 1978)".

(Lenticulo)striate arteries are referred to informally as the arteries of stroke. These arteries arise from the middle cerebral artery and are the most common vessels to be affected by ischemic and hemorrhagic strokes (Thompson and Furlan, 1997; Fewel et al., 2003). There is considerable variation in the distribution of these vessels, but occlusion of the main branch of the lenticulostriate artery (or arteriea) will result in a large ganglionic-capsular infarct and can have considerable functional consequences for the patient (Marinković et al., 2001).

> **TIP!**
>
> Each country and area will have their own sets of criteria for prevention, acute care, dealing with recovery and longer term interventions necessary and the reader should familiarize themselves with local protocols.

6.7.2.1 Internal Carotid Artery Occlusion

Occlusion of the internal carotid artery is a very serious cause of cerebrovascular disease. Its occurrence and incidence tends to be underestimated because it can present symptomatically, or not present at all, i.e.

asymptomatic (Flaherty et al., 2004). For those patients that do show symptoms, a variety of presentations can occur. Mild symptoms like transient unilateral blindness, or amaurosis fugax, may occur when small emboli are found within the ophthalmic branch of the internal carotid artery. Other symptoms of internal carotid artery occlusion include headaches, progressive loss of visual acuity (Klijn et al., 2002) or perhaps even syncope (Kashiwazaki et al., 2005). For those patients that do not exhibit symptoms, the course of the occlusion is relatively benign (Powers et al., 2000). However, if symptoms do present, they tend to have a higher risk for a more catastrophic stroke, or indeed death (Klijn et al., 1997; Sacquegna et al., 1982). In addition to this, occlusion of the internal carotid artery may affect the branches of this vessel i.e. the anterior and middle cerebral vessels. Obviously, the greater the amount of vessels affected will result in a greater territory of the brain having its blood supply stopped, or interrupted. If the occlusion involves many vessels, it may result in the patient having hemianopia (reduced or absent vision) and hemiplegia (paralysis of arm, leg or trunk) on the contralateral side to the occlusion. The patient may also have aphasia (difficulty with comprehension and expression of language) if the dominant side of the brain for language is affected.

6.7.2.2 Anterior Cerebral Artery Occlusion

The anterior cerebral artery supplies the medial side of the frontal and parietal lobes as well as the majority of the corpus callosum and the front portion of the diencephalon. The severity of occlusion of the anterior cerebral artery depends on if the recurrent artery of Heubner (medial lenticulostriate artery) is present. If it is present, it can result in a spastic arm, flaccid leg and very brisk reflexes. If the blockage is in the more proximal segment, this can also result in upper motor neuron pathology of the face. It can also present with anosmia if branches to the olfactory bulb and tract are affected. Micturition can also be affected with an extensive anterior cerebral artery occlusion due to a loss of perineal sensation and inability to control the muscles of the pelvic floor. Apathy may also result if the occlusion affects the blood supply to the frontal lobe, or corpus callosum (Kam and Kim, 2008). Less extensive occlusion, perhaps affecting terminal branches, the lower limb may only be affected. This would present with lack of power and reduced sensation, up-going plantar reflexes and the reflexes to be brisk.

6.7.2.3 Middle Cerebral Artery Occlusion

The most noticeable presentation of occlusion of the middle cerebral artery is paralysis of the lower face and arm on the contralateral side. There will also be a sensory deficit of those areas on the affected side with the paralysis (O'Sullivan and Schmitz, 2007). Obstruction of the cortical branches will result in the patient having monoplegia (paralysis of a single limb, i.e. upper limb) or receptive aphasia. If the central branches are affected, the patient will present with hemiplegia as the internal capsule fibers have been affected. In this instance, speech and language is not affected because the connections of the language centers in the contralateral hemisphere will not suffer.

6.7.2.4 Posterior Cerebral Artery Occlusion

Up to one quarter of ischemic strokes in the UK affects the posterior circulation i.e. the pons, medulla oblongata, cerebellum and the occipital and temporal cortices (Flossmann and Rothwell, 2003). They tend to be more difficult to diagnose and do not fall easily into the previously described F.A.S.T. test typically used for more anterior strokes.

The most common cause for these types of stroke is occlusion or embolism from dissection or atherosclerotic disease of the vertebrobasilar system, or embolism arising from the heart. This would affect the vertebral arteries either in the neck or within the cranial cavity, basilar artery or the posterior cerebral vessels.

Symptoms can be atypical from what may be expected of a stroke and can include a sensation of dizziness, diplopia, homonymous visual field defects, ataxia, dysphagia or dysarthria. If there is a clinical presentation of acute onset crossed defects of the cranial nervous system on one side and contralateral motor and sensory deficits, this is diagnostic of posterior cerebral artery occlusion (Tao et al., 2012).

The acute and longer term management of patients with posterior cerebral artery occlusion is the subject of hot debate currently with large scale trials underway to assess the best methods. Current thoughts in management include modifications of lifestyle factors that may have resulted in the stroke, as well as drug therapies to reduce blood pressure and lipid levels as well as antiplatelet therapy or anticoagulation provided other risk factors have been eliminated (Merwick and Werring, 2014).

6.7.3 Transient Ischemic Attack (TIA)

According to the Royal College of Physicians (UK) National Clinical Guidelines for Stroke (Royal College of Physicians, 2012) document, a TIA is defined as "an acute loss of focal cerebral or ocular function with symptoms lasting less than 24 h and which is thought to be due to inadequate cerebral or ocular blood supply as a result of low blood flow, thrombosis or embolism associated with diseases of the blood vessels, heart, or blood (Hankey and Warlow, 1994)".

6.7.4 Subarachnoid Hemorrhage

This condition results from bleeding from an aneurysm (85%), ruptured blood vessel (10%) or vascular anomaly (5%; van Gijn and Rinkel, 2001). The blood starts to accumulate in the subarachnoid space. It differs from the other cerebrovascular incidents described as the onset is generally sudden, with extremely painful headache and the neurological symptoms are non-specific.

The National Health Service (UK) states that a subarachnoid hemorrhage accounts for about 1 in every 20 strokes (NHS, 2014b, subarachnoid hemorrhage). The majority of these are caused by an aneurysm, but often the cause of is unknown. Indeed, an aneurysm can be completely asymptomatic, with symptoms only present when they rupture.

Severe head injury can also result in a subarachnoid hemorrhage, when it is referred to as a traumatic subarachnoid hemorrhage. In both cases, the blood loss happens in the subarachnoid space resulting in irritation of the meninges, severe headache, neck stiffness and perhaps also a loss of consciousness.

6.7.5 Vascular Dementia

Dementia is a collection of symptoms which can include a loss of memory, change of mood, poor motor control (if caused by a stroke), depression, anxiety, behavioral changes, difficulty with language, thoughts and making decisions. Vascular dementia is the second most common cause of dementia (after Alzheimer's disease, Alzheimer's Society, 2011). The vasculature of the brain is affected by hypertension, hypercholesterolemia and diabetes mellitus.

There are three main types of vascular dementia – stroke related, subcortical and mixed. Stroke related dementia happens due either to thrombotic or ischemic reasons. The degree of how this affects the patient will depend on the areas affected by the stroke. The stroke may affect areas of the brain responsible for motor function of the limbs, or speech production. Multi-infarct dementia is the most common cause, where there are multiple small strokes, or infarcts, affecting the brain.

Subcortical vascular dementia affects the smaller blood vessels deep in the cerebral tissue. Symptoms typical with this disease include lack of coordination walking, difficulty in speech production (apraxia (difficulty producing words) or dysarthria (weakness of speech musculature)) and perhaps also incontinence.

Mixed vascular dementia means that a stroke or subcortical vascular dementia coexists with Alzheimer's disease. As this is a mixed disease, it can exhibit any characteristic of each disease. It affects approximately 10% of those patients with dementia (Alzheimer's Society, 2011).

6.8 OTHER PATHOLOGIES AFFECTING THE CEREBRAL CIRCULATION

6.8.1 Subdural Hemorrhage

The meninges will be dealt with in more detail in Chapter 7. Please refer to this chapter to gain a better understanding of the layers of the meninges, and the location of this type of hemorrhage.

Bleeding of the superior cerebral vein at the point it penetrates the superior sagittal sinus typically causes a subdural hemorrhage. Typically, patients who are affected by this are elderly, or perhaps with a history of alcohol abuse. The brain atrophies in both of these conditions, resulting in stretching of the superior cerebral vein. As only so much pressure can be exerted on the veins, they will eventually rupture. There is no actual space at the interface between the dura and arachnoid. This type of bleed actually creates that space for the blood to collect at that junction (Haines, 2002).

There are three types of onset of a subdural hemorrhage – acute (generally after high impact injury, e.g. road traffic accident); subacute (which takes weeks for signs and symptoms to present); and chronic (typically the elderly population suffer these, and it can take some time to show symptoms and signs).

With an acute subdural heamorrhage, the mortality rates are rather high. The NHS (2013, Subdural Haematoma) states the following:

- "under 40 years old have a 20% risk of dying
- 40–80 years old have a 65% risk of dying
- 80 years old or over have an 88% risk of dying"

The outlook for survival rates from subacute and chronic subdural hematomas are better than the figures quoted for the acute subdural hematoma.

6.8.2 Extradural Haemorrhage

The meninges will be dealt with in more detail in Chapter 7. Please refer to this chapter to gain a better understanding of the layers of the meninges, and the location of this type of hemorrhage.

This type of hemorrhage has been classified previously as due to rupture of the middle meningeal artery (Moore et al., 2006). However, alongside the middle meningeal artery is a pair of dural venous sinuses, which also pass through the foramen spinosum (where the artery enters the cranial cavity), and this arterial origin has now been disputed (Fishpool et al., 2006).

The blood (whether arterial and/or venous in origin) collects in the layer between the outer periosteal layer of dura and the calvaria. Typically, there is trauma (e.g. fracture) at the pterion, where the parietal, frontal, petrous temporal and greater wing of the sphenoid all unites. These patients initially have a lucid interval where no obvious signs are present. This is because it takes time for the blood to accumulate in the extradural space, and affect the cerebral tissue by compression. Increase in blood in the extradural space will then result in a drop in the patient's level of consciousness, as assessed by the internationally recognized classification for consciousness – the Glasgow Coma Scale (Teasdale and Jennett, 1974).

6.8.3 Cavernous Sinus Thrombosis

Infections in the "danger triangle", i.e. the orbit, upper part of the face and the paranasal sinuses can spread deep into the cranial cavity via communications with the superior ophthalmic veins. Infection can pass into the cavernous sinus. The first thing the body does to try to prevent the further spread of this infection is to create a thrombus. If this happens in the cavernous sinus, a cavernous sinus thrombosis can result.

This can affect the structures within the sinus, especially if inflammation also occurs, resulting in thrombophlebitis. The abducent nerve may well be affected, as well as the other cranial nerves within the lateral wall of the venous sinus.

Treatment would be with intravenous antibiotics, and it may be necessary to give steroids to reduce inflammation in and around the cavernous sinus. Anticoagulants may also be used, provided that hemorrhagic complications have been excluded (Migirov et al., 2002). Cavernous sinus thrombosis is a very serious condition with a high mortality rate of approximately 30% with up to half of patients who survive suffering from residual deficits (Yarrington, 1977; Levine et al., 1988). If the sphenoid sinuses are affected, surgical sphenoidotomy may be required (Komatsu et al., 2013).

REFERENCES

Alzheimer's Society. What is Vascular Dementia? http://www.alzheimers.org.uk/site/scripts/documents_info.php?documentID=161 (accessed 19.04.2014).

Creasy, J.L., 2011. Dating Neurological Injury. A Forensic Guide for Radiologists, Other Expert Medical Witnesses and Attorneys. Springer, New York, USA, ISBN 978-1-60761-249-0.

Eastcott, H.H.G., 1994. The beginning of stroke prevention by surgery. Cardiovasc. Surg. 2, 164–169.

Fewel, M.E., Thompson, B.G., Hoff, J.T., 2003. Spontaneous intracerebral hemorrhage: a review. Neurosurg. Focus 15, E1.

Fishpool, S.J.C., Suren, N., Roncaroli, Ellis, H., 2006. Middle meningeal artery haemorrhage: an incorrect name. Clin. Anat. 20 (4), 371–375.

Flaherty, M.L., Flemming, K.D., McClelland, R., Jorgensen, N.W., Brown, Jr., R.D., 2004. Population-based study of symptomatic internal carotid artery occlusion. Incidence and long-term follow-up. Stroke 35 (8), e349–e352, Epub 2004.

Flossmann, E., Rothwell, P.M., 2003. Prognosis of vertebrobasilar transient ischaemic attack and minor stroke. Brain 126, 1940–1954.

Haines, D.E., 2002. Fundamental neuroscience, second ed. Churchill Livingstone, NY, USA.

Hankey, G., Warlow, 1994. Transient ischemic attacks of the brain and eye. WB Suanders, London.

Hines, R.L., Marschall, K.E., 2012. Stoelting's Anethesia and co-existing disease, sixth ed. Elsevier Saunders, Philadelphia, USA, ISBN: 978-1-4557-0082-0.

Johnston, S.C., Nguyen-Huynh, M.N., Schwarz, M.E., Fuller, K., Williams, C.E., Josephson, S.A., Hankey, G.J., Hart, R.G., Levine, S.R., Biller, J., Brown, Jr., R.D., Sacco, R.L., Kappelle, L.J.,

Koudstaal, P.J., Bogousslavsky, J., Caplan, L.R., van Gijn, J., Algra, A., Rothwell, P.M., Adams, H.P., Albers, G.W., 2006. National stroke association guidelines for the management of transient ischemic attacks. Ann. Neurol. 60 (3), 301–313.

Jośe, J., 2009. Stroke in children and young adults, second ed. Elsevier, Philadelphia, USA, ISBN 978-0-7506-7418-8.

Kam, S.Y., Kim, J.S.K., 2008. Anterior cerebral artery infarction. Stroke mecanism and clinical imaging study in 100 patients. Neurology 10;70 (24 Pt 2), 2386–2393, doi: 10.1212/01. PubMed.

Kang, H.-S., Moon, H.H., Kwon, B.J., Kwon O-Ki, Kim, S.H., Chang, K-H., 2005. Evaluation of the lenticulostriate arteries with rotational angiography and 3D reconstruction. AJNR 26, 306–312.

Kashiwazaki, D., Kuroda, S., Terasaka, S., Ishikawa, T., Shichinohe, H., Aoyama, T., Ushikoshi, S., Nunomura, M., Iwasaki, Y., 2005. Carotid occlusive disease presenting with loss of consciousness. No Shinkei Geka, 3329-34.34. PubMed.

Klijn, C.J., Kappelle, L.J., Tulleken, C.A., van Gijn, J., 1997. Symptomatic carotid artery occlusion. A reappraisal of hemodynamic factors. Stroke, 282084-2093.2093. PubMed.

Klijn, C.J., Kappelle, L.J., van Schooneveld, M.J., Hoppenreijs, V.P., Algra, A., Tulleken, C.A., van Gijn, J., 2002. Venous stasis retinopathy in symptomatic carotid artery occlusion: prevalence, cause, and outcome. Stroke, 33695-701.701. PubMed.

Kloppenborg, R.P., Zwanenburg, J.J., Luijten, P.R., Brundel, M., Hendrikse, J., Nederkoorn, P.J., 2011. Imaging of small cerebral blood vessels using 7-Tesla MRI. Ned. Tijdschr. Geneeskd. 155, A3232, PubMed.

Komatsu, H., Matsumoto, F., Kasai, M., Kurano, K., Sasaki, D., Ikeda, K., 2013. Cavernous sinus thrombosis caused by contralateral sphenoid sinusitis: a case report. Head Face Med. 9, 9.

Levine, S.R., Twyman, R.E., Gilman, S. The role of anticoagulation in cavenous sinus thrombosis. Neurol. 38:517-522; 1988

Marinković, S., Gibo, H., Milisavljević, M., Ćetković, M., 2001. Anatomic and clinical correlations of the lenticulostriate arteries. Clin. Anat. 14 (3), 190–195.

Merwick, Á., Werring, D., 2014. Clinical review – posterior circulation ischaemic stroke. BMJ, 348, doi: http://dx.doi.org/10.1136/bmj.g3175.

Migirov, L., Eyal, A., Kronenberg, J., 2002. Treatment of cavernous sinus thrombosis. IMAJ 4, 468–469.

Moonis, G., Granados, A., Simon, S.L., 2002. Epidural hematoma as a complication of sphenoid sinusitis and epidural abscess. A case report and literature review. Clin. Imag. 26, 382–385.

Moore, K.L., Dalley, A.F., Agur, A.M.R., 2006. Clinically Oriented Anatomy, fifth ed. Lippincott Williams and Wilkins, Baltimore, MD.

National Health Service. Cerebrovascular disease. http://www.nhs.uk/conditions/cerebrovascular-disease/Pages/Definition.aspx (accessed 16.05.2014).

National Health Service. Subarachnoid haemorrhage. http://www.nhs.uk/Conditions/Subarachnoid-haemorrhage/Pages/Introduction.aspx (accessed 19.05.2014).

National Health Service. Subdural heamatoma. http://www.nhs.uk/conditions/subdural-haematoma/Pages/Introduction.aspx (accessed 19.05.2014).

National Health Service. Stroke – Act F.A.S.T. (accessed 16.06.2014).

National Institute for Health and Care Excellence CG68 Stroke: algorithm 1. http://www.nice.org.uk/nicemedia/live/12018/41317/41317.pdf (accessed 16.06.2014).

National Institute for Health and Care Excellence CG68 Stroke: algorithm 2 http://www.nice.org.uk/nicemedia/live/12018/41318/41318.pdf (accessed 16.06.2014).

Nor, A.M., Davis, J., Sen, B., Shipsey, D., Louw, S.J., Dyker, A.G., Davis, M., Ford, G.A., 2005. The recognition of stroke in the emergency room (ROSIER) scale: development and validation of a stroke recognition instrument. Lancet Neurol. 4 (11), 727–734.

O'Sullivan, S.B., Schmitz, T.J., 2007. Physical rehabilitation. F a Davis Company, Philadelphia, ISBN 0803612478.

Park, H.K., Bae, H.G., Choi, S.K., Chang, J.C., Cho, S.J., Byun, B.J., Sim, K.B., 2008. Morphological study of sinus flow in the confluence of sinuses. Clin. Anat. 21 (4), 294–300.

Powers, W.J., Derdeyn, C.P., Fritsch, S.M., Carpenter, D.A., Yundt, K.D., Videen, T.O., Grubb, Jr., R.L., 2000. Benign prognosis of never-symptomatic carotid occlusion. Neurology, 54878-882.882. PubMed.

Rajput, A.J., Rozdilsky, B., 1971. Extradural heamatoma following frontal sinusitis. Arch. Otolaryngol. 94, 83–85.

Roos, K.L., Tunkel, A.R., 2010. Bacterial infections of the central nervous system. Elsevier, Amsterdam, Netherlands, ISBN 9780444520159.

Royal College of Physicians. National Clinical Guidelines for Stroke (2012). Prepared by the Intercollegiate Stroke Working Party. UK. http://www.rcplondon.ac.uk/sites/default/files/national-clinical-guidelines-for-stroke-. fourth-edition.pdf (accessed 16.06.2014).

Sacquegna, T., De Carolis, P., Pazzaglia, P., Andreoli, A., Limoni, P., Testa, C., E Lugaresi, E., 1982. The clinical course and prognosis of carotid artery occlusion. J. Neurol. Neurosurg. Psychiatry, 451037-1039.1039. PubMed.

Stroke: Diagnosis and initial management of acute stroke and transient ischaemic attack (TIA). National Institute for Health and Care Excellence CG68 http://www.nice.org.uk/guidance/CG68/chapter/introduction (accessed 08.05.2014).

Symonds, C., 1955. The Circle of Willis. Br. Med. J 1, 119–124.

Tao, W.D., Liu, M., Fisher, M., Wang, D.R., Li, J., Furie, K.L., Hao, Z.L., Lin, S., Zhang, C.F., Zeng, Q.T., Wu., 2012. Posterior versus anterior circulation infarction: how different are the neurological deficits? Stroke 43, 2060–2065.

Teasdale, G., Jennett, B., 1974. Assessment of coma and impaired consciousness. Lancet 304 (7872.), 81–84.

Thompson, D.W., Furlan, A.J., 1997. Clinical epidemiology of stroke. Neurosurg Clin N Am 8, 265–269.

van Gijn, J., Rinkel, G.J.E., 2001. Subarachnoid haemorrhage: diagnosis, causes and management. Brain 124 (2), 249–278.

World Health Organization, 1978. Cerebrovascular diseases: a clinical and research classification. WHO, Geneva.

Wright, S.N., Kochunov, P., Mut, F., Bergamino, M., Brown, K.M., Mazziotta, J.C., Toga, A.W., Cebral, J.R., Ascoli, G.A., 2013. Digital reconstruction and morphometric analysis of human brain arterial vasculature from magnetic resonance angiography. Neuroimage. 2013 Nov 15 82, 170–181, doi: 10.1016/j.neuroimage.2013.05.089. Epub 2013 May 28.

Yarrington, C.T., 1977. Cavernous sinus thrombosis revisited. Proc R Soc Med 70, 456–459.

CHAPTER 7

Essential Anatomy and Function of the Spinal Cord

7.1 MENINGES

The brain and spinal cord are surrounded by layers of tissue called meninges. The meninges are not nervous tissue. It is comprised of three layers – pia mater, arachnoid mater and dura mater, from deep to superficial. Between the arachnoid mater and pia mater is a space called the subarachnoid space, and it is this space that the cerebrospinal fluid circulates.

7.1.1 Meningeal Layer Around the Brain

As like the spinal cord, the brain is surrounded by the meninges and is arranged (from deep to superficial) as the pia, arachnoid and dura mater. Collectively, the pia and arachnoid mater are referred to as the *leptomeninges*.

7.1.1.1 Leptomeninges

The leptomeninges, or the pia and arachnoid mater, are typically described as two separate membranes, but they are joined by connective tissue in the subarachnoid space. The pia mater covers the surface of the brain dipping in between the gyri of the cerebral hemispheres and the folia (equivalent to gyri) of the cerebellum. It is comprised of two types of fibers – elastic and reticular, and is covered by cerebral vessels superficially in the subarachnoid space.

The arachnoid mater surrounds the brain loosely, separated from the dura by a potential subdural space. It dips into the longitudinal fissure, but not into the sulci. Subarachnoid cisternae separate the pia and arachnoid mater at intervals. The cisterna magna (cerebellomedullary cisterna) is continuous below with the subarchnoid space around the spinal cord. The pontine cisterna is found on the anterior aspect of the pons, containing the basilar artery. The chiasmatic and interpeduncular

Essential Clinical Anatomy of the Nervous System. http://dx.doi.org/10.1016/B978-0-12-802030-2.00007-8

cisternae lies superior to the pons, and found between the temporal lobes. The interpeduncular cisterna contains the Circle of Willis. The cisterna of the lateral sulcus is found anterior to the temporal lobes. The cisterna of the great cerebral vein is found between the cerebellum and the corpus callosum (splenium). This contains, as the name suggests, the great cerebral vein and is connected around the brainstem with the interpeduncular cisterna.

The subarachnoid space communicates with the fourth ventricle by apertures (see Chapter 2), and is continuous with the perineural space found around the optic and olfactory nerves. At the point where the dural venous sinuses are found, the arachnoid mater projects as the *arachnoid villi*. It is the arachnoid villi which are responsible for the reabsorption of the cerebrospinal fluid. Visible enlargements of the arachnoid villi are referred to as *arachnoid granulations* and protrude into the dural venous sinuses, most notably into the superior sagittal sinus.

The pia mater on the other hand covers the surface of the brain and passes between the gyri of the cerebral hemispheres, and also the folia of the cerebellum. The pia mater is composed of both elastic and reticular fibers, and is covered superficially by the cerebral vessels which are in the subarachnoid space.

7.1.1.2 Dura Mater

Finally, the outermost layer is the dura mater and surrounds the brain. It is comprised of two layers, though it is difficult to differentiate these with the naked eye. It is comprised of two fused layers – and internal *meningeal layer* and an outer *endosteal layer*. At points along its course, the dural venous sinuses separate it. The inner meningeal layer is limited internally by flattened cells. The external endosteal layer is also referred to as the *endocranium*. This endosteal layer is adherent to the skull, especially at the point of the sutures and also at the base of the skull. At the foramina of the skull and the sutures, the endosteal layer is continuous with the pericranium. Where the endosteal layer is continuous with the foramina, it provides a sheath for the cranial nerves.

The meningeal layer sends four main areas of folds deeper into the cranial cavity. These are the *falx cerebri*, the *tentorium cerebelli*, the *falx cerebelli* and the *diaphragm sellae*.

The falx cerebri is the largest of the folds of dura mater. It attaches anteriorly to the crista galli, and runs posteriorly to fuse with the tentorium cerebelli. In its upper edge, it divides to form the superior sagittal sinus, and its lower edge is where it divides to form the inferior sagittal sinus. The upper edge is attached onto the frontal, parietal and occipital bones in an anterior to posterior direction, respectively. The lower edge of the falx cerebri passes down to separate the left and right cerebral hemispheres from one and other. It typically follows the corpus callosum, linking the two cerebral hemispheres.

The tentorium cerebelli supports the occipital lobes above, and separates them from the cerebellum below. The internal concave surface is free and along with the dorsum sellae of the sphenoid bone, forms the boundary of the tentorial notch, occupied primarily by the midbrain. The outer convex border encloses the transverse sinus posteriorly and is attached to the inner aspect of the skull. More anterior, this margin encloses the superior petrosal sinus and is attached to the upper portion of the petrous temporal bone. Toward the apex of the petrous temporal bone, the two borders of the tentorium cerebelli cross each other. The free border passes above and is anchored to the anterior clinoid process on either side. The attached border passes below and is then attached onto the posterior clinoid process. When the two borders have crossed each other, the triangular area of dura between them forms the roof of the cavernous sinus. The roof is continuous with the diaphragma sellae.

The falx cerebelli lies inferior to the tentorium cerebelli. It superior border is attached to the lower aspect of the tentorium. The posterior border contains the *occipital sinus* and is attached to the occipital bone. The anterior border is free and projects between the hemispheres of the cerebellum.

The diaphragma sellae forms a dural roof for the sella turcica. It covers the *hypophysis*, and has an opening for the *infundibulum*. The optic chiasma lies partly or completely above the diaphragma sellae.

Nerve supply of the dura mater:

(1) Anterior cranial fossa dura – ophthalmic nerve (via the anterior and posterior ethmoidal nerves)
(2) Middle cranial fossa dura – meningeal branches of the maxillary and mandibular nerves (of the trigeminal nerve)

(3) Posterior cranial fossa dura – meningeal branches of the hypoglossal and vagus nerves

(4) Tentorium cerebelli – tentorial branches of the ophthalmic nerve

(5) Falx cerebri – tentorial branches of the ophthalmic nerve

7.1.2 Meningeal Layer Around the Spinal Cord

The three layers of meninges – pia, arachnoid and dura mater, as with the brain, surround the spinal cord.

7.1.2.1 Pia Mater

The pia mater of the spinal cord is comprised of collagen and reticular fibers. The reticular tissue is wrapped closely around the spinal cord and passes posteriorly into the anterior median fissure. The collagen fibers are external to the reticular fibers forming bundles. It is this point where the vessels of the surface of the spinal cord are found. On each side, the collagenous fibers send a thin longitudinal septum laterally. The lateral edge of this *denticulate ligament* is free apart from some processes which fuse with the arachnoid and dura mater, thus helping to anchor the spinal cord.

7.1.2.2 Arachnoid Mater

The arachnoid mater of the spinal cord is delicate. Superiorly, it is continuous with the arachnoid mater within the cranial cavity. Inferiorly, it ends with the dural sac and is pierced by the *filum terminale*, which passes from the *conus medullaris*, or the lower end of the spinal cord. The subarachnoid space is wide with some strands of connective tissue crossing it.

7.1.2.3 Dura Mater

The outer layer of the dura is a very tough fibrous tissue extending from the foramen magnum superiorly to the sacrum and coccyx below. At the level of the foramen magnum, it is continuous with the dura mater from within the cranial cavity. Directly inferior to the foramen magnum, the dura mater is especially thick and highly vascular. In the lower end of the vertebral column, the dura mater narrows, and ends at the level of the second sacral segment. It is prolonged at the filum of the dura mater toward the coccyx, and blends with the periosteum.

The dura mater is separated from the walls of the vertebral canal by the epidural space, containing semifluid fat and many veins. These veins are comparable to the dural venous sinuses of the cranial cavity and lie between the periosteum of the vertebral canal and the dura mater. The epidural fat extends into the intervertebral foramina. The epidural space can be used to inject analgesic agents into for pain relief. This is discussed under the Section 7.3 later in this chapter.

7.2 CLINICAL ASSESSMENT

7.2.1 Lumbar Puncture

As the spinal cord ends at approximately the level of the second lumbar vertebra, but the subarachnoid space continues to the level of the second sacral vertebra, samples of cerebrospinal fluid can be extracted. Prior to the extraction of the cerebrospinal fluid, pressure measurements can be taken. A lumbar puncture can be taken to aid diagnosis of a number of conditions including meningitis, subarachnoid hemorrhage or Guillan–Barré syndrome (immunological condition causing damage to the myelin sheath).

The lumbar puncture is performed with the patient lying on one side with their legs pulled up toward their chest. A local anesthetic is first given to the area to provide pain relief. This results in flexion of the spine and allows the insertion of the needle. The sample of cerebrospinal fluid is taken from below the level of the second lumbar vertebra as the spinal cord terminates at that point. Therefore, obtaining the sample of cerebrospinal fluid from below that point will not place the spinal cord at risk o damage from the needle entering the subarachnoid space. The fluid is collected from this space, and analyzed in the laboratory.

7.3 PATHOLOGIES

7.3.1 Headaches

- *Clinical note*. As the brain itself is insensitive, headaches tend to be of a vascular origin, or of dural origin.

7.3.2 Head Injury

- The dura is supplied by the anterior and posterior meningeal branches of a variety of arteries
- The middle meningeal artery is perhaps the most important branch of the maxillary artery

- In head injuries, trauma to the lateral aspect of the skull, at the pterion (where the greater wing of the sphenoid, squamous part of the temporal bone, parietal bone and frontal bone unite), may result in rupture of the middle meningeal artery
- The middle meningeal artery enters the cranial cavity through the foramen spinosum of the sphenoid bone
- It runs superiorly and laterally for a variable distance in a groove on the squamous part of the temporal bone
- It divides into a frontal and parietal branch, and it is the frontal branch which can be particularly vulnerable
- Laceration of this vessel can occur even without a skull fracture
- Further details can be found in the previous chapter (Chapter 6)

7.3.3 Epidural Analgesia

Epidural analgesia is used in acute pain control and can be used in a variety of clinical situations. It can be used during childbirth or trauma to the chest, abdomen or pelvis in providing effective pain management (Faculty of Pain Medicine, 2010). It involves the insertion of a catheter into the epidural space appropriate to the area to be provided with analgesia. It is essential to follow local guidelines in its insertion, and monitoring of the patient, and also to be aware of any side effects, and risks associated with this procedure. Again, the choice of the analgesic should be dictated by local guidelines.

7.4 SPINAL NERVES

There are 31 pairs of nerves that connect with the spinal cord as the spinal nerves. There are eight in the cervical region, 12 in the thoracic region, five in the lumbar region, five in the sacral region and one coccygeal nerve.

Each spinal nerve has a dorsal and ventral root. It is the dorsal root which carries information related to afferent fibers and the ventral root carries efferent information. In the first cervical and the coccygeal nerves, there is no dorsal root present. The nerve roots pass for a small distance within the dural sac around the spinal cord. The roots then pierce through the dura and enter through the *intervertebral foramina*. At this point, the dorsal root ganglion is also found and it has the cell bodies of the afferent fibers which are about to enter the spinal cord.

Distal to the ganglion, both the ventral and dorsal roots come together to form the common spinal nerve.

The spinal cord, cylindrical in shape as previously described, is larger in the lower cervical segments and in the lumbosacral territories. The lower cervical and first thoracic levels are larger where the brachial plexus arises from, to supply the upper limbs. At the lower end, the lumbar and sacral areas are larger due to the origins of the lumbosacral plexus innervating the lower limbs and pelvic organs.

7.5 SPINAL CORD

7.5.1 White Matter/Gray Matter

The spinal cord is a cylindrical structure which, in the adult, runs from the lower end of the medulla oblongata (with which it is continuous) to the junction between the first and second lumbar vertebrae. In the newborn infant it is slightly longer. It is surrounded by a variety of protective devices.

(1) *Vertebrae*. Just as the skull protects the brain, the spinal cord is surrounded by the vertebral column, made up of a series of individual vertebrae. Whereas the bones of the skull do not move relative to one another (with the exception of the temporomandibular joint), the vertebrae do; this mobility gives flexibility to the trunk. Examine some vertebrae. Despite regional differences, typical vertebrae possess:
    (a) a massive weight-bearing part the body,
    (b) a posterior arch which, with the back of the body, forms the vertebral canal(or foramen) around the spinal cord; and
    (c) three processes which stick out from the vertebra. Two *transverse processes* and a *posterior spinous process*. These are for the attachment of muscles to move the vertebrae.

There is a large fibrous joint between the bodies of adjacent vertebrae – the intervertebral disc – and smaller synovial joints exist between the transverse processes of adjacent vertebrae. In the thoracic region there are extra articulations for ribs, the first and second cervical vertebrae are highly modified and the sacral vertebrae are fused together for strength.

(2) *Meninges and their spaces.* The same three layers of meninges
that surround the brain also surround the spinal cord. Again, the
innermost layer, the pia mater intimately surrounds the spinal
cord, while the outermost layer, the dura mater is partly attached
to the bone. Since the spinal cord ends at L1/2, this is where the
pia ends also (except for a thin projection, the *filum terminale*).
However, the dura and arachnoid continue down to the level of S2;
this means that there is a very large subarachnoid space below the
level of termination of the spinal cord and it is here that lumbar
punctures are performed.

Not only are the meninges themselves protective, but also the fluid-
filled meningeal spaces (such as the subarachnoid space) dampening
down cord movements.

The spinal cord is not uniform in diameter. It is enlarged in the cervical
and in the lumbosacral regions, since these deal with the upper and lower
limbs, respectively. The spinal cord consists of a number of segments (8
cervical, 12 thoracic, 5 lumbar and 5 sacral, plus 2–3 coccygeal). From
each segment, emerge two series of rootlets on each side; a dorsal series
and a ventral series. These come together to form a single dorsal root (on
which is located a swelling, the dorsal root ganglion) and a single ventral
root on each side. These in turn join to form a single spinal nerve on each
side for each segment. These nerves exit through intervertebral foramina.

The dorsal roots/rootlets are associated with incoming (or afferent)
sensory information such as touch, pain, temperature, and proprioception. The cell bodies of these nerves form the DRGs. The ventral roots/
rootlets are associated with outgoing (or efferent) information, i.e. motor commands to muscles. The cell bodies of these nerves are in the ventral gray horn of the cord.

Soon after its formation, a spinal nerve divides into

(1) a posterior primary ramus which supplies skin and muscle behind
the vertebral column, and
(2) an anterior primary ramus which supplies skin and muscle in front
of the vertebral column.

In the case of the nerves supplying the neck, upper limb and lower
limb, the nerves come into close proximity and intermingle as plexuses.

This is often interpreted as a safety device, since each muscle and (more importantly) each movement are now innervated by more than one spinal cord segment.

7.5.1.1 Segmental Distribution of Nerves

No matter how nerve fibers arrive in an area, the region of skin supplied by a particular segment of the spinal cord is called a dermatome. Dermatomes are at best rather approximate, and there is often considerable overlap between adjacent dermatomes. Nevertheless, loss of feeling (anesthesia) in a particular dermatome area could indicate that there was damage or disease of the spinal cord segment (or spinal nerve/roots) supplying that area. Increased or abnormal sensation (often painful) could similarly be evoked, e.g. by a "slipped disc" pressing on nerves/ nerve roots. Altered sensation over the area of distribution of a named peripheral nerve (rather than a dermatome) would suggest damage to a nerve distal to a plexus.

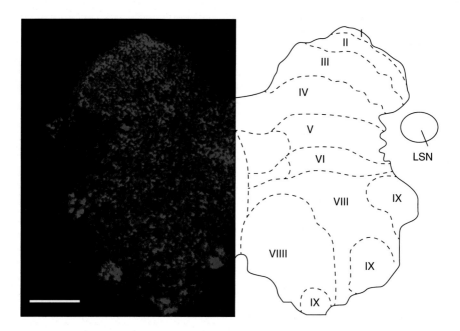

Fig. 7.1. The general layout of the spinal cord. This section is taken at the fifth lumbar vertebral level using immunofluorescence on the left side for a neuronal marker NeuN. The neuronal cells are highlighted in red within the spinal cord section. The diagrammatic representation on the right shows Rexed's laminae (1952) numbered I–X. A small nucleus is also found out in the dorsolateral white matter called the lateral spinal nucleus (LSN). The scale bar represents 500 μm.

When the spinal cord is examined in transverse section, it is composed of a central gray matter (butterfly-shaped) comprising cell columns oriented along the rostro-caudal axis (containing neuronal cell bodies, dendrites and axons that are both myelinated and unmyelinated), surrounded by the white matter comprising the ascending and descending myelinated and unmyelinated fasciculi (tracts). The general layout of the spinal cord is shown in Figure 7.1. Based on detailed studies of neuronal soma size (revealed using the Nissl stain), Rexed (1952) proposed that the spinal gray matter is arranged in the dorso-ventral axis into laminae and designated them into 10 groupings of neurons identified as I–X. Table 7.1 provides an overview of each lamina in terms of the neuronal characteristics, input, output and related functions of each of these territories.

Table 7.1. The Input, Output and Functions of Each of the Laminae Within the Spinal Cord				
Lamina	Characteristics	Input	Output	Function
I	Small neurons and large marginal cells. Fusiform, multipolar, flattened and pyramidal cells. Small dendrites	Fine myelinated and unmyelinated dorsal root fibers	Contralateral spinothalamic tract	Noxious and thermal sensation
II	Stalked and islet cells. Interneurons	Unmyelinated primary afferent input. Very little input from large myelinated primary afferents. Descending dorsolateral fasciculus fibers	Some neurons projecting from the spinal cord (projection neurons), some passing to different laminae and some with axons confined to a lamina in the region of the dendritic tree of that cell, e.g. intralaminar interneurons, local interneurons and Golgi Type II cells	Modulate the activity of nociceptive and temperature afferent fibers
III	Slightly larger cells than lamina II, but with a neurophil similar to that of lamina II. Variable cell size	Hair follicles. Pacinian corpuscles. Rapidly and slowly adapted fibers. Aβ fibers. Dendrites of cells from laminae III-VI	Propriospinal and interneurons	Light tactile stimulation and strong pressure
IV	Variety of cell types	Collaterals and from large primary afferent fibers Input also arises from the substantia gelatinosa (lamina II)	Projection to the ipsilateral and contralateral spinothalamic tracts. Spinocervical, spinocerebellar and spinohypothalamic tracts. Nucleus accumbens and the lateral parabrachial area and reticular nuclei	Pain, temperature and crude touch. Light mechanical stimulation

Table 7.1. The Input, Output and Functions of Each of the Laminae Within the Spinal Cord *(cont.)*

Lamina	Characteristics	Input	Output	Function
V	Thick nerve fibers. Comprised of a lateral and medial zone. Triangular, star shaped and spindle cells. Difficult to distinguish lamina V from lamina VI.	Large primary afferent collaterals. Monosynaptic information from Aβ, Aδ and C fibers. Descending fibers from the corticospinal and rubrospinal tracts	Brainstem and thalamus via ipsilateral and contralateral spinothalamic tracts. Cerebellum, dorsal column nuclei, reticular formation, rostromedial medulla. Local projections (more typical of the lateral area)	Multireceptive. Nociception and mechanorecpetion
VI	Broad layer. Developed most in the cervical and lumbar enlargements. Variety of cell shapes here	Joint and muscle spindle afferents. Rubrospinal and corticospinal pathways (in its lateral portion)	Spinocerebellar pathway	Integration of somatic motor processes. Reflex pathways via interneurons
VII	Renshaw cells. Triangular, fusiform and multipolar cell types found here	Tendon and muscle afferents. Laminae II–VI. Sympathetic preganglionic neurons constituting the intermediolateral cell column in the thoracolumbar (T1–L2/3) and the parasympathetic neurons located in the lateral aspect of the sacral cord (S2–4), i.e. visceral afferent fibers	Innervates postganglionic cells in the sympathetic or parasympathetic ganglia. Inhibitory interneurons which synapse on the alpha motor neurons	Autonomic nervous system. Inhibitory interneurons which synapse on the alpha motor neurons
VIII	Shape varies across the spinal cord	Vestibulospinal and reticulospinal fibers. Descending motor tracts from the cerebral cortex and the brainstem	Motor neurons innervating the intrafusal fibers	Modulate motor activity
IX	Large and smaller motor neurons. Size and shape varies across the spinal cord	Descending motor tracts from the cerebral cortex and the brainstem	Motor neurons	Extrafusal skeletal muscle fibers. Intrafusal muscle fibers. Nociceptive
X	Decussation of axons. Small to medium sized cells. Triangular, star shaped and spindle cells	Small myelinated and unmyelinated fibers. Convergence of somatic and visceral primary afferent input	Lateral parabrachial nucleus, amygdala and the nucleus of the solitary tract, medullary and pontine reticular formation	Nociceptive and mechanoreceptive information

REFERENCES

Best practice in the management of epidural analgesia in the hospital setting (2010). Faculty of Pain Medicine of the Royal College of Anaesthetists. http://www.rcoa.ac.uk/system/files/FPM-EpAnalg2010_1.pdf (accessed 02.08.2014).

Rexed, B., 1952. The cytoarchitectonic organisation of the spinal cord in the cat. J. Comp. Neurol. 96, 414–495.

CHAPTER 8

Spinal Tracts – Ascending/Sensory Pathways

8.1 GENERAL INTRODUCTION

The ascending tracts carry sensory information up to the brain. Many of these pathways start with *"spino"*, meaning that the tract begins in the spinal cord, and will ascend to various brain regions. The latter half of the name of the pathway will give where the pathway terminates.

Therefore, the following summary can be used.

(1) Spinocerebellar tract – sensory tract carrying information from the *spinal cord* to the *cerebellum*
(2) Spinothalamic tract – sensory tract carrying information from the *spinal cord* to the *thalamus*
(3) Spinohypothalamic tract – sensory tract carrying information from the *spinal cord* to the *hypothalamus*
(4) Spinoreticular tract – sensory tract carrying information from the *spinal cord* to the *reticular formation (in brainstem)*
(5) Spinotectal (spinomesencephalic) tract – sensory tract carrying information from the *spinal cord* to the *periaqueductal gray matter, superior colliculus and a variety of nuclei located in the reticular formation*
(6) Spino-olivary tract – sensory tract carrying information from the *spinal cord* to the *olivary nuclei (superior and inferior) in the medulla*

8.2 DORSAL COLUMN/MEDIAL LEMNISCUS

8.2.1 Anatomical Location and Function

One of the ascending pathways does not have "spino" in its name, but is called the *dorsal column*. It is also referred to as the *medial lemniscus*, and is part of the ascending pathway found in the *dorsal white matter*.

Essential Clinical Anatomy of the Nervous System. http://dx.doi.org/10.1016/B978-0-12-802030-2.00008-X

Functionally it is comprised of two parts, generally from mid-thoracic level upwards in the spinal cord:

(1) Cuneate fasciculus
     This pathway transmits information from the *upper limbs, neck* and *upper trunk*, i.e. superior to approximately the sixth thoracic vertebral level (T6).

(2) Gracile fasciculus
     This pathway transmits information from the *lower limbs* and *lower trunk*, i.e. inferior to approximately the sixth thoracic vertebral level.

The *cuneate fasciculus* is located *medially* in the spinal cord as it ascends from the spinal cord to the brain, with the *gracile fasciculus* more *lateral*.

Specifically, the medial lemniscus conveys information about several functions as detailed below in Table 8.1

8.2.2 Location of Neuronal Cell Bodies

The medial lemniscus has a three-neuron pathway to its final point of termination. From the periphery, the medial lemniscus conveys information to the *primary somesthetic cortex*, also known as the *primary somatosensory cortex*. This is located in the *postcentral gyrus*.

The receptors for the gracile and cuneate fasciculi arise from *Pacinian corpuscles, Meissner's corpuscles, tendon organs* and *muscle spindles*.

Tables 8.2 and 8.3 below highlight the location of each of the three neurons in the medial lemniscus tract for the cuneate and gracile fasciculi, respectively.

Table 8.1. The Functions Associated with the Medial Lemniscus
Functions of the medial lemniscus
Fine touch
Two-point discrimination (discriminative touch)
Pressure
Proprioception (awareness of relative position of our limbs relative to one and other. This comes from our skeletal muscles and joints)
Vibration

Table 8.2. The Location of the Cell Body, Point of Termination of Fibers, and Site of Crossover are Given for the Cuneate Fasciculus

Cuneate Fasciculus

	Cell Body	Termination	Ipsilateral/Contralateral to input (site of crossover)
First order	Dorsal root ganglia superior to T6	Caudate medulla – cuneate nucleus	Ipsilateral
Second order	Cuneate nucleus	Ventral poterolateral thalamic nucleus	Cross – over here (prior to medial lemniscus)
Third order	Ventral posterolateral thalamic nucleus	Postcentral gryus (primary somesthetic cortex; passing through the internal capsule)	Contralateral

Table 8.3. The Location of the Cell Body, Point of Termination of Fibers, and Site of Crossover are Given for the Gracile Fasciculus

Gracile Fasciculus

	Cell Body	Termination	Ipsilateral/Contralateral to input (site of crossover)
First order	Dorsal root ganglia inferior to T6	Caudate medulla – gracile nucleus	Ipsilateral
Second order	Gracile nucleus	Ventral poterolateral thalamic nucleus	Cross – over here (prior to medial lemniscus)
Third order	Ventral posterolateral thalamic nucleus	Paracentral lobule (primary somesthetic cortex; passing through the internal capsule)	Contralateral

The general sensory fibers from the head and neck are primarily conducted in the trigeminal nerve and will be dealt with in Chapter 10.

8.2.3 Clinical Assessment
8.2.3.1 Fine Touch

TIP!

When examining a patient's sensory system, do not provide suggestions to them as it may influence their interpretation of the examination. DO NOT say to the patient if they notice any changes during the examination, as this suggests that they should expect to notice a change.

8.2.3.1.1 Light Touch Examination

The purpose of this is to serve as an introductory examination to identify any areas or regions where there may be pathologies present. Simple testing should first be done and as the extremities are easy to access without causing too much discomfort to the patient, these should be examined first.

> **TIP!**
>
> Testing of *light touch sensation* to identify any areas of abnormality MUST be done when the patient's eyes are closed. This ensures that they do not anticipate where the stimulus is applied. Equally, you MUST tell the patient what you are going to do before doing it. This helps build trust with the patient but also allows them to be fully informed about what you are doing, and that they consent to it.

(1) Introduce yourself to the patient stating who you are, and in what capacity and grade you are functioning, e.g. student, physician, surgeon, therapist etc.

(2) Advise the patient that you want to test for sensation in their extremities initially when their eyes are closed.

(3) Tell them that you will touch them with either cotton wool, or using a specialist Von Frey filament on various regions of the arms, hands, legs and feet on both the left and right hand sides. You do not need to use the term Von Frey but inform them that it is material made of nylon that allows you to test sensation.

(4) Ask the patient if they have understood what you have said and answer any questions they may have.

(5) Now, ask the patient to close their eyes and to say "yes" every time they feel you touch, and also to report any feelings of discomfort, pain or abnormal sensations.

(6) Working systematically with the cotton wool or Von Frey fibers, touch all dermatome regions of the upper and lower limbs on both the right and left sides of the limbs.

(7) Try to compare both the left and right hand sides of each dermatome region.

> **TIP!**
>
> A dermatome is a region of skin supplied by one single spinal nerve. It allows you to determine what spinal level pathology may be at. The distribution of our dermatomes can be slightly variable but can give an indication at what level approximately the pathology may exist, e.g. upper cervical, lower thoracic and so on.

If you identify a level, or levels, where there appears to be an alteration in sensation, identify exactly at what point this occurs. When recording the findings from the clinical examination, it can be used on the American Spinal Injury Association's (ASIA, 2014) worksheet produced as the International Standards for Neurological Classification of Spinal Cord Injury (ISNCSCI).

8.2.3.2 International Standards for Neurological Classification of Spinal Cord Injury (ISNCSCI)

This sheet allows for assessment and classification of motor and sensory function and is classified as follows.

Sensory assessment: Light touch and pinprick are assessed separately and a score out of 2 is given for each dermatome on the right and left side of the body.

The dermatomes that are assessed are cervical (C2, C3, C4, C5, C6, C7, C8), thoracic (T1, T2, T3, T4, T5, T6, T7, T8, T9, T10, T11, T12), lumbar (L1, L2, L3, L4, L5) and sacral (S1, S2, S3, S4, S5).

For sensation the following scoring system is used.

Sensory Score	Classification
0	Absent
1	Altered
2	Normal
NT	Not testable

A score is given for light touch right (LTR, totaling 56, i.e. 2 for each of the vertebral levels stated above) and light touch left (LTL, totaling 56, i.e. 2 for each of the vertebral levels stated above). This is recorded as follows:

$$LTR + LTL = out\ of\ 112$$

A score is also given for pinprick right (PPR, totaling 56, i.e. 2 for each of the vertebral levels stated above) and pinprick left (PPL, totaling 56, i.e. 2 for each of the vertebral levels stated above). This is recorded as follows:

$$PPR + PPL = out\ of\ 112$$

Motor assessment: Each major muscle group is assessed in turn. These are (with vertebral level being assessed):

Upper extremity (assess left and right):

- C5 – Elbow flexors
- C6 – Wrist extensors
- C7 – Elbow extensors
- C8 – Finger flexors
- T1 – Finger abductors

Lower extremity (assess left and right):

- L2 – Hip flexors
- L3 – Knee extensors
- L4 – Ankle dorsiflexors
- L5 – Long toe extensors
- S1 – Ankle plantar flexors

For each muscle group, on the left and right hand sides, a score of 5 is given for each on its functions based on the following classification detailed below.

Motor Score	Classification of Muscle Function
0	Total paralysis
1	Palpable or visible contraction
2	Active movement, with gravity eliminated
3	Active movement against gravity
4	Active movement, against some resistance
5	Active movement, against full resistance
5*	Normal but corrected for pain/disuse
NT	Not testable

Therefore in assessment in a patient without pathology of the musculature and its innervation, the following formula would result in 50/50 for the upper extremities and 50/50 for the lower extremities in total, i.e.

Upper Extremities Right (UER) 25 (out of a max. of 25)

+

Upper Extremities Left (UEL) 25 (out of a max. of 25)

=

$$\underline{50}$$

Lower Extremities Right (UER) 25 (out of a max. of 25)

+

Lower Extremities Left (UEL) 25 (out of a max. of 25)

=

$$\underline{50}$$

8.2.3.2.1 Two-point Discrimination

The two-point discrimination test is used to assess if the patient is able to identify two close points on a small area of skin, and how fine the ability to discriminate this are. It is a measure of tactile agnosia, or the inability to recognize these two points despite intact cutaneous sensation and proprioception. It generally would represent a brain injury but can be done alongside assessment of dermatomes for light touch (as with medial lemniscus / dorsal column testing) or pain testing (as with the spinothalamic tract (see Section 8.4.1).

This test can be done using calipers. It used to be said that the examiner could improvise by using a clean paperclip which has been unraveled to create two separate points, of relatively close distance apart. The normal distance should be approximately 5mm apart. However, this has not been found to be the best to use in a clinical setting, with other commercial products now available (Skirven et al., 2011).

In addition, it has been found not to be the best in producing consistent and repeatable findings on examination (Bell-Krotoski and Buford, 1997). It is however, still preferred by hand surgeons as an assessment of nervous innervation of the hand (Moberg, 1962; Dellon, 1978; MacKinnon and Dellon, 1985). More recently, however, it has been discussed that it may serve a purpose if an aesthesiometer is used by "experts" and using a robust scale like the one formed by the Inflammatory Neuropathy Cause and Treatment (INCAT) group (van Nes et al., 2008) using the INCAT Sensory Sumscore (ISS; Merkies et al., 2000). That being said, it can still be quite inconsistent and depends on many variables in its interpretation and should be used as an indicator and not entirely prescriptive of two-point discrimination.

8.2.3.2.2 Romberg's Test

The function of a patient's dorsal column is assessed using Romberg's test. There are several inputs that the patient must have to be able to do this test:

(1) Intact visual input
(2) Intact vestibular function
(3) Proprioception

Generally, at least two of these are needed to undertake this test. The following instructs how to conduct the Romberg's test.

TIP!

Always inform the patient what you are going to be doing before instructing them. This helps build trust and allows for fully informed consent.

The examination should consist of the following.

(1) Ask the patient to stand up with their feet together, arms by their side and eyes open.
(2) Then, ask the patient to close their eyes for approximately 20–30 s.
(3) The patient may exhibit mild swaying which is normal.
(4) It is possible to repeat the test two times to help assessment.
(5) If the patient loses their balance, it is said that they have a positive Romberg's test, or Romberg's sign.

TIP!

Stand close to the patient in case they lose their balance, and prevent them from falling if they are not able to undertake this test successfully.

The reasoning behind this test is in the fact that one of the three features of balance has been removed, i.e. visual input. If the patient shows a positive Romberg's sign/test, there is either a problem with the vestibular apparatus or the patient has a proprioceptive disorder.

Proprioception can also be tested of the area under examination by movement of joints.

(1) First, distal joints can be assessed, and then move centrally. The interphalangeal joints can be assessed initially
(2) Move the joint upwards and downwards, with their eyes open, telling the patient what each movement is.
(3) Then, ask the patient to close their eyes, and move the joints up and down, asking what the patient feels, i.e. if the joint was moved upwards or downwards.
(4) If abnormalities are found, move centrally to other joints, e.g. wrist, elbow, etc.

TIP!

Vary the movements upward and downward, i.e. to prevent the patient predicting any movement. If the same sequence is followed, the patient may try to predict what *should* be expected of them.

8.2.3.3 Fine Touch
Fibers for light touch will travel in the dorsal columns ipsilaterally, but also simple tactile information for pain will travel contralaterally via the spinothalamic tract (Section 8.4.1). Fine touch can be tested, simply by using cotton wool. The skin should be *touched* (i.e. do not run the fibers over the skin) by the cotton wool over the various dermatomes relevant to the area being examined. The patient must be told of what you are going to do first, and that their eyes *must* be closed during that test.

8.2.3.4 Vibration Sense
A 128 Hz tuning for should be used. The patient should close their eyes and the vibrating tuning fork can be placed on a joint. The examiner then should stop the instrument from vibrating. The patient should be instructed to tell the examiner when the tuning fork *stops* vibrating.

8.2.4 Pathologies
(1) Unilateral interruption of the spinal cord
   A unilateral lesion of the *dorsal column*, or at the level *before they decussate*, will result in an interruption of the sensory information *below the level of the pathology*. It will affect the *same side of the body as the lesion*.
   If there is an interruption of the pathways *above* the level of decussation (i.e. at the medial lemnsicus or above), it will result in loss of sensory input *below the level of the pathology* on the *opposite side of the body*.

Specifically, the sensations which will be lost are those that are found in the dorsal column/medial lemniscal pathway. Therefore, it will result in a loss of information related to fine touch, two-point discrimination, pressure and proprioception.

(2) Tabes dorsalis

This condition is also known as *syphilitic myelopathy*. It is due to infection by syphilis and affects the large diameter fibers and their related ganglia that enter the dorsal columns.

This progressive condition occurs over many years and affects the sensory information in the dorsal columns, as previously described. The patient will have a variety of symptoms including weakness, abnormalities of sensation (paresthesia or reduced/absent sensation) and a positive Romberg's test. Muscle weakness can also be found due to loss of the sensory input of the reflex arc of the stretch reflex. Once nerve damage occurs, it cannot be reversed.

(3) Brown-Séquard Syndrome

This is a rare condition which occurs when there is a hemisection of the cord affecting either side (Laporte, 2006). It may result from compression by a tumor, ischemia or trauma. There are two key features of this syndrome:

(a) IPSILATERAL TO LESION – Loss of fine touch, vibration sense and proprioception. This is due to interruption of the dorsal column / medial lemniscus pathway. There will be spastic weakness/paralysis also on the affected side due to interruption of the motor neuron pathway conducted via the lateral corticopsinal tract (Chapter 9, Section 9.1).

(b) CONTRALATERAL TO LESION – Loss of temperature and pain sensation due to interruption of the spinothalamic tract (Section 8.4.1).

(4) Subacute combined degeneration of the spinal cord

This is also known as Lichtheim's disease and is degeneration of the spinal cord at the level of the dorsal columns, but also the lateral columns (and therefore affecting the spinothalamic tracts too. It is commonly caused by a deficiency of vitamin B12, vitamin E or copper. It presents in a variety of ways typically resulting in weakness and paresthesia ("pins and needles") in the limbs and trunk. Treatment is targeted at first identifying the cause and treating that deficiency. For example, a chronic alcoholic may have reduced levels of vitamin B12 and folic acid. Vitamin b12 should be given in that case either to prevent the subacute combined degeneration of the spinal cord from happening, or to reduce its effects.

8.3 SPINOCEREBELLAR TRACT

8.3.1 Anatomical Location and Function

The spinocerebellar tract occupies a large portion of the white matter on its outer aspects. It spans across the dorsal, lateral and ventral areas of the peripheral aspect of the white matter. The spinocerebellar tract is comprised of four key territories, dependent on where the information is arising from, as it enters the spinal cord to then ascend to the cerebellum (Table 8.4). The four key areas of the spinocerebellar tract are:

(1) Dorsal (posterior) spinocerebellar tract

This pathway transmits information from the caudal aspect of the body and legs functionally, although its associated nucleus (**Clarke's nucleus**) receives information from all parts of the body from C8 to L2. The information transmitted to the cerebellum within this pathway is something that you are unaware of. Specifically the information will relate to both individual muscles, and groups of them. It is referred to as *non-conscious proprioception*. It allows the cerebellum to coordinate posture and the movement of the *lower limb musculature*. It is an *ipsilateral* pathway.

(2) Cuneocerebellar tract

The cuneocerebellar tract is similar to the dorsal spinocerebellar tract. Specifically, it conveys information related to the *upper limbs*. As Clarke's nucleus is not present above the level of C8, the fibers entering form the upper limb pass to the medulla via the *cuneate fasciculus*, synapsing in the *accessory cuneate nucleus* before passing to the cerebellum. Again, this pathway conveys *non-conscious proprioceptive* information from muscle spindles and Golgi tendon organs from the upper limb musculature. It is an *ipsilateral* pathway.

Table 8.4. The Functions Associated With the Spinocerebellar Tract
Functions of the spinocerebellar tract
Transmission of proprioceptive information (from joints, muscles and tendons)
Transmission of tactile information (i.e. from skin)

(3) Ventral (anterior) spinocerebellar tract
Information conveyed in the ventral spinocerebellar tract arises from Golgi tendon organs at the junction between the tendon and the muscle of the *lower limbs*. Initially the information passes from one side of the body then crosses over at the spinal cord ascending in the ventral spinocerebellar tract. At the level of the pons, these fibers then crossover again back to the same side the information had arisen from in the *superior cerebellar peduncle*. This then passes to the cerebellum. This tract conveys information about movement of the entire limb and adjustments to the posture. The information terminates on the *ipsilateral* side but crosses *twice*, first at the level of the spinal cord and second at the superior cerebellar peduncle.

(4) Rostral spinocerebellar tract
The rostral spinocerebellar tract is like the ventral spinocerebellar tract but the difference is that it conveys information about the upper limbs from the Golgi tendon organs. It is an ipsilateral pathway and the information passes to the cerebellum via the *inferior cerebellar peduncles*.

8.3.2 Location of Neuronal Cell Bodies

The spinocerebellar tracts have a two-neuron pathway to its final point of termination. From the periphery, the spinocerebellar tracts convey information to the *cerebellum*, also known as the *primary somatosensory cortex*. This is located in the *postcentral gyrus*.

The receptors for the spinocerebellar tracts arise from *muscle spindles* and *Golgi tendon organs*.

The tables below (Tables 8.5–8.8) below highlight the location of each of the two neurons in the four spinocerebellar tracts.

Table 8.5. The Location of the Cell Body, Point of Termination of Fibers, and Site of Crossover are Given for the Dorsal Spinocerebellar Tract			
Dorsal Spinocerebellar Tract			
	Cell Body	Termination	Ipsilateral/Contralateral to input (site of crossover)
First order	Dorsal root ganglia of C8 – L2 fibers (from muscle spindles and, to a lesser extent, Golgi tendon organs)	Clarke's nucleus	Ipsilateral
Second order	Clarke's nucleus	Cerebellum	Ipsilateral

Table 8.6. The Location of the Cell Body, Point of Termination of Fibers, and Site of Crossover are Given for the Cuneocerebellar Tract

Cuneocerebellar tract

	Cell Body	Termination	Ipsilateral/Contralateral to input (site of crossover)
First order	Dorsal root ganglia of the upper limb (from Golgi tendon organs and muscle spindles)	Accessory cuneate nucleus	Ipsilateral
Second order	Accessory cuneate nucleus	Cerebellum	Ipsilateral

Table 8.7. The Location of the Cell Body, Point of Termination of Fibers, and Site of Crossover are Given for the Ventral Spinocerebellar Tract

Ventral Spinocerebellar Tract

	Cell Body	Termination	Ipsilateral/Contralateral to input (site of crossover)
First order	Dorsal root ganglia of lower limbs (of Golgi tendon organs at the musculotendinous junction)	Intermediate gray matter	Crossover in the spinal cord – contralateral to input
Second order	Intermediate gray matter	Cerebellum	Crossover in the pons (after the superior cerebellar peduncle) – to ipsilateral side of input

Table 8.8. The Location of the Cell Body, Point of Termination of Fibers, and Site of Crossover are Given for the Rostral Spinocerebellar Tract

Dorsal Spinocerebellar Tract

	Cell Body	Termination	Ipsilateral/Contralateral to input (site of crossover)
First order	Dorsal root ganglia of upper limbs (from Golgi tendon organs)	Intermediate gray matter	Crossover in the spinal cord
Second order	Intermediate gray matter	Cerebellum	Crossover in the pons (after the superior cerebellar peduncle)

8.3.3 Clinical Assessment

The assessment of the cerebellum can be undertaken by assessing the neurological system in full (Section 8.2) and by also examining specifically the cerebellum. There are several aspects to examining the integrity of the cerebellum.

(1) Always introduce yourself to the patient (in any clinical examination or history taking) and state your position

(2) *Assess gait*. Ask the patient to walk from one side of the room (or examining area) to the other. If they normally use an aid to walking, they should be allowed to do so.

(3) *Heel to toe*. The patient should be asked to walk forward by placing one heel in front of the toes then switching to the opposite side and to keep walking in this fashion for a short distance

(4) Romberg's test. Further details are also found in Section 8.2.
   (a) Ask the patient to stand up with their feet together, arms by their side and eyes open.
   (b) Then, ask the patient to close their eyes for approximately 20–30 s.
   (c) The patient may exhibit mild swaying which is normal.
   (d) It is possible to repeat the test two times to help assessment. If the patient loses their balance, it is said that they have a positive Romberg's test, or Romberg's sign.

(5) Check for a *resting tremor* by having the patient place their arms and hands out straight.

(6) *Assess muscle tone and power* as discussed in detail in Section 8.2

(7) *Check for dysdiadochokinesis*. Ask the patient to touch one dorsal surface of the hand with the palmar surface of the opposite hand. The opposite hand should then rotate to the dorsal surface of the opposite hand. This alternating palmar/dorsal surface onto the opposite hand should be repeatedly as rapidly as possible for the patient. Dysdiadochokinesis is the inability to undertake this rapid movement.

(8) *Finger to nose*. The patient should touch their nose then the examiners finger which is held in space. The examiner should move their examining finger and the patient should repeat the movement of touching their nose and the moving examiners finger.

(9) *Heel to shin test*. The patient should be asked to place the heel of one foot at the knee of the opposite leg. Then roll the heel down the front of the shin and back up. This should be repeated several times. Repeat this on the opposite side several times too.

8.4 ANTEROLATERAL SYSTEM

The anterolateral system is a series of pathways that conveys information related to temperature and pain. It also transmits information related to tactile sensations like light touch and firm pressure. Crude touch is also transmitted in this pathway, which is where the individual knows that they have been touched, but unable to localize it to an exact point. This system contrasts that of fine touch which is transmitted within the dorsal column/medial lemniscus (see Section 8.2).

8.4.1 Spinothalamic Tract
8.4.1.1 Ventral/Lateral Spinothalamic Tract
8.4.1.1.1 Anatomical Location and Function

The spinothalamic tract is comprised of two separate components that transmit different information – the ventral (or anterior) and lateral spinothalamic tracts. The ventral spinothalamic tract transmits information related to crude touch and firm pressure, whereas the lateral spinothalamic tract transmits information related to temperature and pain.

Information related to *pain, temperature, crude touch* and *firm pressure* would enter the dorsal horn from the lower and upper limbs. Those fibers related to pressure and touch will enter through the dorsal rootlets medial division. Fibers containing information related to temperature and pain will enter through the *dorsolateral tract of Lissauer* (also called the dorsolateral fasciculus of Lissauer, tract of Lissaeur, or simply the dorsolateral tract).

Those fibers, which enter the dorsolateral tract, pass to the substantia gelatinosa (see Chapter 7). The sensory information arriving in the substantia gelatinosa is modified, and has many interneurons at that point (Gurdt and Perl, 2002; Hantman et al., 2004; Yasaka et al., 2007, 2010).

From here, the fibers then pass through the white matter through the ventral white commissure ascending to the ventral posterior nucleus of the thalamus. They then pass to the postcentral gyrus.

8.4.1.1.2 Location of Neuronal Cell Bodies

The spinothalamic tracts have a three-neuron pathway to its final point of termination. From the periphery, the spinothalamic tracts convey information to the *postcentral gyrus*, (*primary somatosensory cortex*).

The receptors for the spinothalamic tracts arise from the following, depending on what they convey.

- Pain (nociception) – unencapsulated axon endings of A fibers (Aδ) and unmyelinated C fibers (carried in the *lateral spinothalamic tract*)
- Temperature – free nerve endings (carried in the lateral spinothalamic tract)
- Light touch – nerve endings (unencapsulated), *Merkel nerve endings* and *Meissner's corpuscles* (carried in the *ventral spinothalamic tract*)
- Firm pressure – *Ruffini endings* (found in the deep skin; carried in the *ventral spinothalamic tract*)

Table 8.9. This Table Summarizes Where the Cell Bodies are Found of the Spinothalamic Tract and Where Each of the Neurons Terminate, and Where They Crossover, If Applicable

Spinothalamic Tract			
	Cell Body	Termination	Ipsilateral/Contralateral to input (site of crossover)
First order	Dorsal root ganglia	*Substantia gelatinosa* (lamina II) and *nucleus proprius* (laminae III, IV and V)	*Ipsilateral*
Second order	*Substantia gelatinosa* and *nucleus proprius*	*Ventral thalamic nuclei*	*Crossover here* (via the ventral white commissure)
Third order	*Ventral thalamic nuclei*	*Primary somesthetic cortex, cingulate cortex, insular cortex*	*Contralateral*

Table 8.9 above highlights the location of each of the three neurons in the spinothalamic tracts, both lateral and ventral. The only difference is really where the information arises from, and what type of information is conveyed in each, which is described above.

8.4.1.2 Clinical Assessment

The first thing to do when examining the spinothalamic tract, and to determine if there are any deficiencies of it, is to undertake a screening test. This has to be done as a patient's ability to determine sensation is completely dependent on their interpretation of it, and relies completely on their input. Take for example a patient with a psychiatric illness. It may be difficult to ascertain with certainty what they feel and their interpretation of it.

TIP!

When examining a patient's sensory system, do not provide suggestions to them as it may influence their interpretation of the examination. DO NOT say to the patient if they notice any changes during the examination, as this suggests that they should expect to notice a change. With this test, the pinprick test is used to assess perception of pain. *Warn the patient about this before undertaking this examination.* You could also test for light touch sensation along with this examination for an assessment of the dorsal column/medial lemniscus pathways too (see Section 8.2).

8.4.1.2.1 Screening Examination

The purpose of this is to serve as an introductory examination to identify any areas or regions where there may be pathologies present. Simple testing should first be done and as the extremities are easy to access without causing too much discomfort to the patient, these should be examined first.

> **TIP!**
>
> Testing of *pain sensation* to identify any areas of abnormality MUST be done when the patient's eyes are closed. This ensures that they do not anticipate where the stimulus is applied. Equally, you MUST tell the patient what you are going to do before doing it. This helps build trust with the patient but also allows them to be fully informed about what you are doing, and that they consent to it.

> **TIP!**
>
> When testing pain sensation using a pin (or clean cocktail stick) ALWAYS ensure that you use only one pin per patient to avoid cross-contamination.

(1) Introduce yourself to the patient stating who you are, and in what capacity and grade you are functioning, e.g. student, physician, surgeon, therapist, etc.

(2) Advise the patient that you want to test for sensation in their extremities initially when their eyes are closed.

(3) First do a test on the sternum of the patient with the pin when the patient's eyes are open to allow them to understand what the sensation looks and feels like.

(4) You can say to the patient that that sensation would be 10 out of 10.

(5) Tell them that you will touch them with a small pin (*though will not penetrate the skin*) on various regions of the arms, hands, legs and feet on both the left and right hand sides.

(6) Ask the patient if they have understood what you have said and answer any questions they may have.

(7) Now, ask the patient to close their eyes and to say "yes" every time they feel you touch, and also to report any feelings of discomfort, pain or abnormal sensations.

(8) Working systematically with the pin, touch all dermatome regions of the upper and lower limbs on both the right and left sides of the limbs.

(9) Try to compare both the left and right hand sides of each dermatome region.

If you identify a level, or levels, where there appears to be an alteration in sensation, identify exactly at what point this occurs, or there is an alteration compared to the control of 10/10 on the sternum as tested on initial examination.

When recording the findings from the clinical examination, it can be used on the American Spinal Injury Association's (ASIA) worksheet produced as the International Standards for Neurological Classification of Spinal Cord Injury (ISNCSCI), as previously described.

8.4.1.2.2 Further Testing

For a complete assessment of all dermatome regions, it will be necessary to check the chest, abdomen and pelvis to complete the ASIA's sheet on ISNCSI.

In addition to this, you can also assess test for temperature sensation, though this is not commonly tested for. If assessment of the spinothalamic tract for temperature sensation is required, exactly the same regions are tested in the dermatome map as previously described for light touch and pinprick sensation on the ASIA's sheet on the ISNCSCI.

A test tube filled with hot water and also cold water can be used on the patient's skin. Again, it is essential to tell the patient what you will be doing and also to make sure the water is neither too hot nor too cold as to cause damage or burns either to the examiner or the patient's skin. This is generally tested when there is an abnormality to pain perception.

If an abnormality is detected when examining pinprick sensation, *nerve conduction studies* can be used as more advanced testing of the patient's spinothalamic tract and nerve conduction. Specialist neurological advice should be sought if there is any uncertainty, or if nerve conduction studies are considered appropriate. It may well be that magnetic resonance imaging (MRI) may be appropriate dependent on the clinical presentation of the patient, and the findings on examination.

8.4.1.3 Pathologies

First it is wise to consider the definitions of abnormalities of touch and pain.

Abnormalities of *touch* can be defined as follows.

(1) Anesthesia
    This is defined as a loss of sensation of pressure, touch, pain or temperature
(2) Hypoesthesia
    This is defined as less than "normal" feeling to any sensory modality

(3) Hyperesthesia
This is defined as an exaggerated or increased feeling to any
sensory modality

Abnormalities of *pain* can be defined as follows.

(1) Analgesia
This is defined as the absence of pain in the conscious patient
(2) Hypoalgesia
This is defined as a reduced sensation to pain
(3) Hyperalgesia
This is defined as an increased sensation of pain

The following demonstrates some causes for abnormalities in sensation of pain, temperature, crude touch and firm pressure.

(1) Diabetes mellitus
Diabetes mellitus can result in peripheral neuropathy causing
damage to the nerves that feed into the spinothalamic tract,
resulting in an inability to feel pain in those nerves, with
paresthesia, hypoalgesia or analgesia. This would be referred to as
a *sensory neuropathy*.
(2) Syringomyelia
Cavitation of the spinal cord due to a cyst, or syrinx (hence the
name of this condition) can be congenital or acquired. The most
common cause of syringomyelia is due to a congenital cause –
Arnold–Chiari formation, where the cerebellum pushes downwards
and the syrinx develops at the cervical level. Acquired causes can be
due to trauma (e.g. hemorrhage), inflammation (as in meningitis)
or tumors. The syrinx can enlarge and press on the ventral white
commissure – the area where the spinothalamic tract crosses over
from the ipsilateral side of input to the contralateral side to ascend
to the thalamus. When the syrinx extends to the ventral region, it
will interrupt these fibers and result in a *bilateral loss of temperature
and pain sensations*. This is because it will affect both left and right
sides where the fibers decussate in the ventral white commissure.
It will affect the level of the spinal cord where the syrinx is found.
(3) Spinal cord injury
If a spinal cord injury is present, whatever that cause may be
(e.g. tumor, trauma), it can affect both the dorsal column/

medial lemniscus and spinothalamic tracts. If that is the case, a *unilateral lesion* will result in *loss of proprioception, fine touch, pressure* and *vibration on the affected side below the level of the lesion*. However, there will be reduced perception of *pain* below the level of the lesion on the *opposite side of the lesion*. This is because the spinothalamic tract (carrying information for pain and temperature) crosses the midline to ascend to the thalamus. This is referred to as the Brown–Séquard syndrome, as previously described in Section 8.2.

(4) Anterolateral cordotomy

This procedure is one that is not routinely carried out and is reserved for those patients who have terminal cancer, or another incurable disease leading to intractable pain.

(5) Lateral medullary syndrome

This is also referred to as Wallenberg's syndrome. The injury happens to the lateral part of the medulla and affects the spinothalamic tract at that point as it ascends on its way to the primary somesthetic cortex. It can be caused by blockage of either the posterior inferior cerebellar artery or a branch from it causing an infarct in the lateral medulla. It results in a variety of presentations affecting the facial, glossopharyngeal, vagus and accessory nerves (Rea, 2014). It also results in sensory defects across the trunk and limbs on the opposite side of the body to the lesion. Specifically, there will be a loss of *temperature* and *pain* sensation on the *opposite* side of the body. In addition, and there will be *loss of temperature* and *pain sensation* on the same side of the body as the lesion.

(6) Anterior spinal artery syndrome

This syndrome is due to occlusion of the artery that supplies the ventral aspect of the spinal cord – the anterior spinal artery. Occlusion of this vessel will result in an interruption of the blood supply to the ventral portion of the spinal cord. It will result in the patient having potentially extensive motor paralysis below the level of the lesion. As it can affect the ventral white commissure, where the spinothalamic tract crosses to the opposite side, there will be a bilateral loss of pain and temperature sensation at, and below, the level of the affected territory/territories.

8.4.2 Spinohypothalamic Tract

The hypothalamus has a variety of roles in the control of autonomic and neuroendocrine functions. The hypothalamus controls the activity of the viscera and is the main effector of the limbic system, therefore has a control over emotions too. It is complex and has both endocrine and neural components, exerting itself through the circulatory system and the nervous system.

The following table (Table 8.10) again summarizes the functions of the hypothalamus.

In addition, the hypothalamus has three main nuclei, each serving different roles. The main nuclei are the anterior, intermediate and posterior nuclei.

- Anterior nuclei – sleep
  - Circadian rhythm
  - Thirst and water balance

Table 8.10. The Functions of Each of the Hypothalamic Nuclei	
Hypothalamus	
Region and related nuclei	**Functions**
Anterior medial	
Supraoptic nucleus	Release of antidiuretic hormone (vasopressin)
Medial preoptic nucleus	Releases gonadotrophin releasing hormone (GnRH)
Anterior hypothalamic nucleus	Sweating, temperature regulation
Suprachiasmatic nucleus	Maintenance of the circadian rhythm
Paraventricular nucleus	Release of oxytocin, corticotrophin releasing hormone, thyrotrophin releasing hormone and somatostatin
Anterior lateral	
Lateral nucleus	Hunger and thirst
Lateral preoptic nucleus	Maintenance of body temperature
Posterior lateral	
Lateral nucleus	Hunger and consciousness
Posterior medial	
Posterior nucleus	Shivering Release of ADH Raising of blood pressure Mydriasis
Mamillary nuclei	Recollection memory

- Heat loss
- Uterine contraction and ejection of milk
- Intermediate nuclei – emotions
  - Feeding
  - Endocrine activities
- Posterior nuclei – memory
  - Analgesia
  - Conservation of heat
  - Arousal

The afferent information to the hypothalamus is wide and varied. It arises from afferents from the soma and viscera, olfaction, frontal lobe, hippocampus, amygdaloid complex and some of the thalamic nuclei (Snell, 2009). Efferent connections from the hypothalamus include the parasympathetic and sympathetic outflows (in the craniosacral and thoracolumbar regions, respectively), reticular formation, limbic system and the thalamus (anterior nucleus).

In addition, the hypothalamus can be viewed simply as its medial and lateral portions. Stimulation of the lateral hypothalamus causes the parasympathetic outflow to predominate (Milam et al., 1980; Yoshimatsu et al., 1984), and as it has one of the largest descending inputs to the periaqueductal gray (PAG) of the rat, is implicated in descending modulation of spinal neuronal activity, especially that resulting from noxious stimulation, and without affecting reactions to other stimuli (Beitz, 1982; Basbaum and Fields, 1984; Jensen and Yaksh, 1984; Aimone and Gebhart, 1987; Tasker et al., 1987; Aimone et al., 1988). Stimulation of the medial hypothalamus and specifically the ventromedial hypothalamic nucleus, results in domination of the sympathetic outflow (Inoue et al., 1977; Niijima et al., 1984; Yoshimatsu et al., 1984; Saito et al., 1989; Uyama et al., 2004). In addition, the medial hypothalamus (especially the ventromedial area) has been suggested to have an additional role in the motivational reaction to a noxious stimulus (Bester et al., 1995; Braz et al., 2005). Specifically, it has been suggested that it is involved in processing information that may threaten the animal, and organizes the execution of innate defensive behaviors (Siegel, 2005; Borszcz, 2006). The paraventricular area on the other hand consists of several nuclei and controls the autonomic nervous system, regulation of visceral organs (Kannan et al., 1987;

Uyama et al., 2004) and coordinates neurosecretions influencing the pituitary gland (Freund-Mercier et al., 1981).

Previously, a direct spinohypothalamic pathway was not identified (Bowsher, 1957; Mehler et al., 1960; Boivie, 1979; Craig and Burton, 1985) and it was believed that the afferent pathway for somatosensory information to the hypothalamus was transmitted exclusively via indirect, multisynaptic projections. The earliest suggestion of a direct projection from the spinal cord to the hypothalamus was based on anatomical studies in the monkey. Chang and Ruch (1949) demonstrated that sectioning the monkey spinal cord resulted in degeneration at the supraoptic decussation at several levels of the hypothalamus bilaterally. Since then numerous anatomical and electrophysiological studies have revealed that somatosensory and visceral information can reach the hypothalamus through monosynaptic pathways that originate in medullary dorsal horn neurons and from all levels of the spinal cord (Burstein et al., 1987; Katter et al., 1996a,b; Kostarczyk et al., 1997; Zhang et al., 1999; Malick et al., 2000). The physiological studies in the cervical (Dado et al., 1994), thoracic (Zhang et al., 2002) and lumbosacral segments (Burstein et al., 1987, 1991) have shown that the majority of spinal cord neurons projecting to the hypothalamus are strongly activated by noxious thermal and mechanical stimuli with sacral segments activated by noxious stimulation of both visceral and cutaneous structures (Katter et al., 1996a,b).

In addition to physiological studies, Burstein et al. (1987) performed several retrograde tracing techniques by injecting Fluoro-Gold (FG) into the hypothalamus and demonstrated a large number of labeled neurons bilaterally throughout the length of the spinal cord, with approximately half located in the lateral reticulated area and a lesser proportion around the central canal and marginal zone. From their retrograde tracing studies, they also showed a relatively large number of cells in the contralateral superficial dorsal horn, though mainly in the lower cervical cord of the rat. Kayalioglu et al. (1999) injected FG into the rat hypothalamus and from careful observations of their diagrams, demonstrated not only labeled neurons in the deeper laminae and the area around the central canal, but also in lamina I. However, the numbers in lamina I were considerably less than in the area around the central canal. This contrasts with anterograde studies done by Gauriau and Bernard (2004), who found that most of the projecting neurons to the hypothalamus were located in the deeper laminae, with most of these being in the lateral reticulated area of lamina V. However, their study was limited to the cervical segments of the rat. However, what has not been demonstrated clearly is this pathway in humans.

8.4.2.1 Visceral Sensations

It is worthwhile noting here about visceral sensations. These sensations related perhaps to feeling "full" within the bowel and bladder, heartburn or feeling hungry are all vague symptoms and tend to be difficult to localize. There are many pathways related to these routes, with many synapses occurring before reaching the cerebral cortex, which also tend to be under-represented compared to other sensations.

During surgery perhaps of the bowel, bladder, heart and other parts of the gastrointestinal tract, true sensations are not interpreted despite handling of these organs by the surgeon. However, if there is over stretching of these organs or extreme contractions of the bowel, pain may be felt in the organ or organs, which would be true visceral pain, or may be referred to another somatic area. This would be referred to as referred pain.

With referred pain, there may be pathology within the viscera, but which is referred to related areas of the skin. Most of the fibers from the viscera will travel with the sympathetic nerves (originating from the first thoracic vertebra to approximately the second lumbar vertebra). Although the pain may not appear related cutaneously to the visceral organs, they are part of the same vertebral level.

Generally, it is believed that the afferents from the viscera arrive at the spinal cord where the somatic afferents arrive in. It is thought that if there is a large input from the visceral area concerned, the spinothalamic tract neurons will also be activated, hence the feeling of "visceral pain".

In addition, the hypothalamus controls the activity of the visceral organs and is the main effector of the limbic system. Therefore, the hypothalamus exerts its functions through both the nervous system connections, but also from the endocrine portion.

8.4.3 Spinoreticular Tract

The spinoreticular tract has its neurons within the intermediate gray matter and the deep dorsal horn. One group passes to the medullary reticular formation and the other to the pontine reticular formation.

The pathways are both unilateral and contralateral in their projections. Therefore, these pathways transmit sensory information to the reticular formation. From here, this information will be transmitted to the cerebral cortex via the thalamus and upwards to the cerebral cortex.

8.4.4 Spinotectal (Spinomesencephalic) Tract

The neurons of the spinomesencephalic tract are found in the intermediate gray matter and the deep dorsal horn. The spinomesencephalic tract then passes from these spinal cord layers to the midbrain and terminate within the periaqueductal gray matter. It is thought that this tract is then conveyed to the amygdala via the parabrachial nucleus.

8.4.5 Spino-olivary Tract

The spino-olivary tract neurons carry proprioceptive information from the tendons and muscles. The axons will enter the dorsal aspect of the spinal cord and terminate within the dorsal gray matter. From here, they synapse and cross the midline to ascend in the white matter at the junction between the ventral and lateral white columns as the spino-olivary tract. They will terminate on the inferior olivary nucleus. From here, there is a further crossover entering the cerebellum via the inferior cerebellar peduncle. Due to the crossover from one side to the other and back again, they convey information from the ipsilateral side related to cutaneous as well as proprioceptive information.

REFERENCES

Aimone, L.D., Bauer, C.A., Gebhart, G.F., 1988. Brain-stem relays mediating stimulation-produced anti-nociception from the lateral hypothalamus in the rat. J. Neurosci. 8 (7), 2652–2663.

Aimone, L.D., Gebhart, G.F., 1987. Spinal monoamine mediation of stimulation-produced anti-nociception from the lateral hypothalamus. Brain Res. 403, 290–300.

American Spinal Injury Association's International Standards for Neurological Classification of Spinal Cord Injury. Found at: http://www.asia-spinalinjury.org/elearning/ASIA_ISCOS_high.pdf (accessed 10.06.2014).

Basbaum, A.I., Fields, H.L., 1984. Endogenous pain control systems: brainstem spinal pathways and endorphin circuitry. Annu. Rev. Neurosci. 7, 308–338.

Beitz, A.J., 1982. The organisation of afferent projections to the midbrain periaqueductal grey of the rat. Neuroscience 7, 133–159.

Bell-Krotoski, A., Buford, Jr., W., 1997. The force/time relationship of clinically used sensory testing instruments. J. Hand Ther. 10 (4), 297–309.

Bester, H., Menendez, L., Besson, J.M., Bernard, J.F., 1995. Spino(trigemino)parabrachiohypothalamic pathway: electrophysiological evidence for an involvement in pain processes. J. Neurophysiol. 73, 568–585.

Boivie, J., 1979. An anatomical reinvestigation of the termination of the spinothalamic tract in the monkey. J. Comp. Neurol. 186, 343–370.

Borszcz, G.S., 2006. Contribution of the ventromedial hypothalamus to generation of the affective dimension of pain. Pain 123, 155–168.

Bowsher, D., 1957. Termination of the central pain pathway in man: the conscious appreciation of pain. Brain 80, 606–621.

Braz, J.M., Nassar, M.A., Wood, J.N., Basbaum, A.I., 2005. Parallel "pain" pathways arise from subpopulations of primary afferent nociceptor. Neuron 47, 787–793.

Burstein, R., Cliffer, K.D., Giesler, Jr., G.J., 1987. Direct somatosensory projections from the spinal cord to the hypothalamus and telencephalon. J. Neurosci. 7, 4159–4164.

Burstein, R., Dado, R.J., Giesler, Jr., G.J., 1991. Physiological characterization of spinohypothalamic tract neurons in the lumbar enlargement of rats. J. Neurophysiol. 66, 261–284.

Chang, H.-T., Ruch, T.C., 1949. Spinal origin of the ventral supraoptic decussation (Gudden's commissure) in the spider monkey. J. Anat. 83, 1–9.

Craig, A.D., Burton, H., 1985. The distribution and topographical organisation in the thalamus of anterogradely-transported horseradish peroxidase after spinal injections in the cat and raccoon. Exp. Brain Res. 58, 227–254.

Dado, R.J., Katter, J.T., Giesler, Jr., G.J., 1994. Spinothalamic and spinohypothalamic tract neurons in the cervical enlargement of rats. I. Locations of antidromically identified axons in the cervical cord white matter. J. Neurophysiol. 71, 1003–1021.

Dellon, A.L., 1978. The moving two-point discrimination test: clinical evaluation of the quickly adapting fiber/receptor system. J Hand Surg. 3, 474–481.

Freund-Mercier, M.J., Stoeckel, M.E., Moos, F., Porte, A., Richard, P., 1981. Ultrastructural study of electrophysiologically identified neurons in the paraventricular nucleus of the rat. Cell Tissue Res. 216, 503–512.

Gauriau, C., Bernard, J.-F., 2004. A comparative reappraisal of projections from the superficial laminae of the dorsal horn in the rat: the forebrain. J. Comp. Neurol. 468, 24–56.

Gurdt, T.J., Perl, E.R., 2002. Correlations between neuronal morphology and electrophysiological features in the rodent superficial dorsal horn. J. Physiol. 540, 189–207.

Hantman, A.W., van der Pol, A.N., Perl, E.R., 2004. Morphological and physiological features of a set of spinal substantia gelatinosa neurons defined by green fluorescent protein expression. J. Neurosci. 24, 836–842.

Inoue, S., Campfield, L.A., Bray, G.A., 1977. Comparison of metabolic alterations in hypothalamic and high fat diet-induced obesity. Am. J. Physiol. 233, R162–168.

International Standards for Neurological Classification of Spinal Cord Injury (ISNCSCI). American Spinal Injury Association. http://www.asia-spinalinjury.org/elearning/ASIA_ISCOS_high.pdf (accessed 28.05.2014).

Jensen, T.S., Yaksh, T.L., 1984. Spinal monoamine and opiate systems partly mediate the antinociceptive effects produced by glutamate at brainstem sites. Brain Res. 321, 287–297.

Kannan, H., Niijima, A., Yamashita, H., 1987. Inhibition of renal sympathetic nerve activity by electrical stimulation of the hypothalamic paraventricular nucleus in anaesthetised rats. J. Autonom. Nerv. Syst. 21, 83–86.

Katter, J.T., Ddo, R.J., Kostarczyk, E., Giesler, Jr., G.J., 1996a. Spinothalamic and spinohypothalamic tract neurons in the sacral spinal cord of rats. I. Locations of antidromically identified axons in the cervical cord and diencephalon. J. Neurophysiol. 75, 2581–2605.

Katter, J.T., Ddo, R.J., Kostarczyk, E., Giesler, Jr., G.J., 1996b. Spinothalamic and spinohypothalamic tract neurons in the sacral spinal cord of rats. II. Responses to cutaneous and visceral stimuli. J. Neurophysiol. 75, 2606–2628.

Kayalioglu, G., Robertson, B., Kristensson, K., Grant, G., 1999. Nitric oxide synthase and interferon-γ receptor immunoreactivities in relation to ascending spinal pathways to thalamus, hypothalamus, and the periaqueductal grey in the rat. Somatosens. Motor Res. 16, 280–290.

Kostarczyk, E., Zhang, X., Giesler, Jr., G.J., 1997. Spinohypothalamic tract neurons in the cervical enlargement of rats: locations of antidromically identified ascending axons and their collateral branches in the contralateral brain. J. Neurophysiol. 77, 435–451.

Laporte, Y., 2006. Charles-Edouard Brown-Sequard: an eventful life and a significant contribution to the study of the nervous system. C R Biol. 329 (5–6), 363–368.

MacKinnon, S.E., Dellon, A.L., 1985. Two-point discrimination tester. J Hand Surg. 10A, 906–907.

Malick, A., Strassman, A.M., Burstein, R., 2000. Trigeminohypothalamic and reticulohypothalamic tract neurons in the upper cervical spinal cord and caudal medulla of the rat. J. Neurophysiol. 84, 2078–2112.

Mehler, W.R., Feferman, M.E., Nauta, W.J.H., 1960. Ascending axon degeneration following anterolateral cordotomy. An experimental study in the monkey. Brain 83, 718–750.

Merkies, I.S., Schmitz, P.I., van der Meche, F.G., van Doom, P.A., 2000. Psychometric evaluation of a new sensory scale in immune-mediated polyneuropathies. Inflammatory Neuropathy Cause and Treatment (INCAT) Group. Neurology 54, 943–949.

Milam, K.M., Stern, J.S., Storlein, L.H., Keesey, R.E., 1980. Effect of lateral hypothalamic lesions on regulation of body weight and adiposity in rats. Am. J. Physiol. 239, R337–343.

Moberg, E., 1962. Criticism and study of methods for examining sensibility in the hand. Neurology 12, 8–19.

Niijima, A., Rohner-Jeanrenaud, F., Jean-Renaud, B., 1984. Role of ventromedial hypothalamus on sympathetic efferents of brown adipose tissue.

Rea, P., 2014. Clinical Anatomy of the Cranial Nerves. Academic Press, San Diego, USA, ISBN 978-0-12-800898-0.

Saito, M., Minokoshi, Y., Shimazu, T., 1989. Accelerated norepinephrine turnover in peripheral tissues after ventromedial hypothalamic stimulation in rats. Brain Res. 481, 298–303.

Siegel, A., 2005. The Neurobiology of Aggression and Rage. CRC Press, Boca Raton.

Skirven, T.M., Osterman, A.L., Fedorczyk, J.M., Amadio, P.C., 2011. Rehabilitation of the Hand and Upper Extremity, sixth ed. Mosby, ISBN: 9780323056021.

Snell, R.S., 2009. Clinical Neuroanatomy, seventh ed. Lippincott Williams and Wilkins, ISBN-10: 0781794277.

Tasker, R.A.R., Choinière, M., Libman, S.M., Melzack, R., 1987. Analgesia produced by injection of lidocaine into the lateral hypothalamus. Pain 31, 237–248.

Uyama, N., Geerts, A., Reynaert, 2004. Neural connections between the hypothalamus and the liver. Anat. Rec. 280A, 808–820.

van Nes, S.I., Faber, C.G., Hamers, R.M., Harschnitz, O., Bakkers, M., Hermans, M.C., Meijer, R.J., van Doorn, P.A., Merkies, I.S., PeriNomS Study Group, 2008. Revising two-point discrimination assessment in normal aging and in patients with polyneuropathies. J. Neurol. Neurosurg. Psychiatry 79 (7), 832–834.

Yasaka, T., Kato, G., Furue, H., Rashid, M.H., Sonohata, M., Tamae, A., Murata, Y., Masuko, S., Yoshimura, M., 2007. Cell-type-specific excitatory and inhibitory circuits involving primary afferents in the substantia gelatinosa of the rat spinal dorsal horn in vitro. J. Physiol. 581, 603–618.

Yasaka, T., Tiong, S.Y.X., Hughes, D.I., Riddell, J.S., Todd, A.J., 2010. Populations of inhibitory and excitatory interneurons in lamina II of the adult rat spinal dorsal horn revealed by a combined electrophysiological and anatomical approach. Pain 151 (2), 475–488.

Yoshimatsu, H., Niijima, A., Oomura, Y., Yamabe, K., Katafuchi, T., 1984. Effects of hypothalamic lesion on pancreatic autonomic nerve activity in the rat. Brain Res. 303, 147–152.

Zhang, X., Wenk, H.N., Gokin, A.P., Honda, C.N., Giesler, Jr., G.J., 1999. Physiological studies of spinohypothalamic tract neurons in the lumbar enlargement of monkeys. J. Neurophysiol. 82, 1054–1058.

Zhang, X., Gokin, A.P., Giesler, Jr., G.J., 2002. Responses of spinohypothalamic tract neurons in the thoracic spinal cord of rats to somatic stimuli and to graded distension of the bile duct. Somatosens. Motor Res. 19, 5–17.

CHAPTER 9

Spinal Tracts – Descending/Motor Pathways

9.1 PYRAMIDAL TRACTS

The pyramidal tracts are comprised of the corticospinal and corticobulbar tracts. These are called as pyramidal tracts as they crossover at the level of the pyramids in the medulla. They are collections of upper motor neuron fibers which go to the spinal cord (corticospinal) or the brainstem (corticobulbar) and control the motor function of the body.

The corticospinal tract is comprised of a ventral and lateral tract. Many of the fibers crossover at the level of the medulla controlling muscles (and therefore movement) of the opposite side of the body.

9.1.1 Ventral and Lateral Corticospinal Tract

The following summary points will highlight the key facts in relation to the ventral and lateral corticospinal tracts. This will allow quick and easy access to the key information relevant to these pathways.

- The origin of the corticospinal tracts is within the primary motor, premotor and supplementary motor cortices, as well as the cingulate motor region within the frontal lobe
- Integration of information occurs with relays between the association fibers from other motor areas. These receive information from the prefrontal, temporal and parietal lobes
- The corticospinal tracts then pass into the *internal capsule* at its posterior limb
- It then passes into the *basis pedunculi* of the midbrain
- The corticospinal tracts then pass to the pons
- Following breaking up into fasciculi and reassembling, they then create the pyramidal tract at a prominence at the medulla
- Crossover occurs at the caudal end of the *medulla*
- Also called *decussation of the pyramids*

Essential Clinical Anatomy of the Nervous System. http://dx.doi.org/10.1016/B978-0-12-802030-2.00009-1

- The majority of fibers (approximately 85%) then pass to the *dorsal half of the lateral funiculus*. These fibers form the *lateral corticospinal tract*
- The remaining 15% of fibers will then form the ventral corticospinal tract. These fibers pass in the medial portion of the ventral funiculus
- In the spinal cord, most axons of the corticospinal tract terminate in the intermediate gray matter and the ventral horn
- The corticospinal tract influences motor neurons only through interneurons within the spinal gray matter

9.2 CLINICAL ASSESSMENT

Clinical assessment of the corticospinal tract should involve a comprehensive neurological examination. As the corticospinal tract helps influence motor neurons, the musculature of the body can be focused on, but should not be done in isolation. The presentation and history will direct toward the most relevant clinical examinations to be undertaken, and to the appropriate investigations. In terms of the motor aspect of the neurological examination, this would include assessment of muscle tone and power (as described in Chapter 8), observation of the muscle groups to identify fasciculations and tendon reflexes. The tendon reflexes to be examined are the jaw, biceps, triceps, knee, ankle and plantar reflexes.

9.3 PATHOLOGIES

9.3.1 Upper Motor Neuron Lesion

If the corticospinal tract is damaged anywhere along its course, from its origins in the cerebral cortex to its termination in the spinal cord, it is referred to as an upper motor neuron lesion. An upper motor neuron lesion is interruption of these pathways, for whatever reason, before the ventral horn neurons, i.e. motor neurons.

The following features will be present in a neurological examination of a patient with an upper motor neuron lesion,

(1) Weak or absent voluntary movement of the muscles
(2) Increased muscle tone called spasticity

(3) *Babinski sign*. Normally stroking the lateral aspect of the sole of the foot will result in plantar flexion of the great toe. With the Babinski sign, the great toe dorsiflexes.

(4) Suppression of the superficial reflexes, i.e. abdominal and cremasteric reflexes. Normally, stroking of the abdomen will result in contraction of the abdominal musculature. With an upper motor neuron lesion, this will be suppressed. Similarly, if the medial side of the thigh is touched gently, the ipsilateral testis will withdraw. In an upper motor neuron lesion, this will also be suppressed in the male.

(5) Clonus may be present. This is alternating contraction and relaxation of muscles when a tendon is stretched.

(6) If the lesion is at the level of the cerebral cortex, the extensor muscles of the lower limb are stimulated by the intact vestibulospinal tract (see Section 9.6.1). This will result in the limb being internally rotated and extended.

(7) If the lesion is at the level of the spinal cord, all paralyzed limbs will be in flexion as the vestibulospinal tracts will be transected.

(8) Facial muscles will be paralyzed only in the lower half of the face due to alternative bilateral pathways at the brainstem.

(9) Atrophy is not a typical feature

TIP!

The Babinski sign is seen in children under the age of 12 months, and does not become completely dorsiflexion until the eighteenth month. It is at this point that there is complete myelination of all of the fibers of the corticospinal tracts.

Causes of an upper motor neuron lesion can be divided up into cerebral, brainstem or spinal cord lesions (Kahan et al., 2009). The more common causes of an upper motor neuron lesion are as follows.

(1) *Stroke*. This can be due to hemorrhage, thrombosis or an embolism. This can present acutely, and should be assessed as detailed in Chapter 6.

(2) *Traumatic brain injury*. This can be due to rapid acceleration or deceleration, impact from blast or from penetration by a foreign body. Traumatic brain injury is assessed based on the anatomical features, severity and mechanism (Saatman et al., 2008).

(3) Multiple sclerosis
(4) Cerebral palsy

9.4 PATHOLOGIES

A wide variety of pathologies can affect the descending pathways, and also many other pathways which have been previously described, and will be described later. The following represent some major diseases which affect parts of the nervous system and will be examined here.

9.4.1 Multiple Sclerosis

Multiple sclerosis (MS) is a condition which causes demyelination of the myelin sheaths which surround the nerves. It also can result in the formation of lesions within the nervous system, also with inflammation, The exact cause of MS is still unknown but has been thought of as due to a variety of influences including genetics and infection (Compston and Coles, 2008).

Within the UK, there are more than 100,000 people with MS, and women are more affected than men, usually presenting at approximately 20–40 years old (MS Trust, 2014). It presents with a wide variety of signs and symptoms from muscular weakness and spasms to visual problems and bowel and bladder symptoms.

Multiple sclerosis generally commences as isolated neurological signs and symptoms over several days. Some patients may experience either motor or sensory symptoms, or both. One of the most obvious presentations can be optic neuritis. A variety of presentations in disturbed vision can occur from blurring to blindness. Most cases resolve spontaneously (Rea, 2014).

Two patterns exist of MS – one where there are symptoms which then improve, or a gradual worsening of the condition where there are no periods of recovery. It has been noted that the occurrence of the symptoms generally peaks during the winter and summer periods (Tataru et al., 2006), but can also be brought on by stress (Heesen et al., 2007). Interestingly, MS is most common in Europe and Scotland; UK has the highest rates in the world (Rothwell and Charlton, 1998; Visser et al., 2012). Visser et al. (2012) suggested that this increase in prevalence in Scotland, especially in Orkney was due to a higher sex ratio and an influence of genes and the environment interacting, though a definitive cause has not been clearly established.

As mentioned, the exact cause of MS has not been identified but it is thought to be of an immunological basis interacting with other factors like genetics and the environmental influence. The main features of this disease are the formation of lesions or plaques within the CNS, an inflammatory component and also the demyelination of the sheaths around the axons.

The plaques within the CNS develop in the white matter of the optic nerve (hence development of optic neuritis), brainstem and within the spinal cord. Specifically, the oligodendrocytes that produce the myelin sheaths are lost, therefore affecting nerve conduction. The peripheral nervous system tends not to be affected with MS. The inflammation which occurs is related to T cells which attack the myelin sheaths, though the precise details, again, or not fully understood (Compston and Coles, 2008).

The progression of MS has been variable, but an attempt to classify the progression of this disease has been created by the following criteria (Lublin and Reingold, 1996).

(1) Relapsing remitting
(2) Secondary progressive
(3) Primary progressive
(4) Progressive relapsing

A variety of treatment options exist, but local policy and procedures in and out from the neurological team should be sought if there is a clinical suspicion of MS. Typically, interferons are the first line of treatment, and steroids may help acute relapses (Tsang and Macdonell, 2011). However, long-term treatments still do not exist at the moment.

9.4.2 Cerebral Palsy

Cerebral palsy is defined as a permanent disorder which generally occurs in fetal life, at birth or shortly after birth affecting the movement or the posture of the patient. It has a wide variety of presentations which can be mild to more severe disability affecting the patient. The patient with cerebral palsy can have spastic paresis or paralysis which affects the descending motor pathways. It may affect cognition and can also present with learning disabilities, involuntary movements and a reduced muscle mass. It can also affect the patient's speech and ability to communicate effectively.

Cerebral palsy can occur due to problems in utero, at birth, or within the first few years after birth, though most occur at birth (Yarnell, 2013). A significant proportion of babies born prematurely will suffer from cerebral palsy, again due to problems at birth. In addition, other issues which arise at birth can increase the chances of someone being born with cerebral palsy like a low birth weight, meconium in the newborns lungs or the delivery method, e.g. caesarean sectioning or birth asphyxia.

Cerebral palsy is assessed on the child's general motor functions, and the earlier this is assessed the better. If further investigations are needed, it may include imaging of the nervous system by either CT or MRI scanning.

There are several types of classification systems used in cerebral palsy, and depend on the area of the clinician's specialty. However, broadly, they are divided up as follows (cerebralpalsy.org)

(1) Severity
(2) Topographical distribution
(3) Motor functioning
(4) Gross motor functioning

However, recently orthopedic surgeons also developed a simpler classification based on the following three categories:

(1) Spastic
(2) Ataxic
(3) Dyskinetic

This system tried to attempt to classify the cerebral palsy into the areas that were affected. The spastic classification affected the cerebral cortex, and was by far the most common type. Ataxic cerebral palsy referred to pathology within the cerebellum, with dyskinetic, or athetoid cerebral palsy showing a pattern of mixed tone within the muscles, and typically affected the basal ganglia and the pyramidal and/or extrapyramidal tracts, which will have been discussed throughout this book.

Management of patient's with cerebral palsy generally involves a multidisciplinary team approach depending on the patient's needs. They may need orthopedic intervention for skeletal abnormalities, pain management teams for the pain of contractures, neurological input for any issues with the nervous system, especially involving continence. Urological input may be necessary for surgical intervention for these issues with incontinence or abnormal bowel or bladder function or medical therapy for provision of anticonvulsants, botulinum toxin or diazepam. Treatment for this condition is based on the symptoms of the patient and is not curative of the underlying causes of this condition. Other members of the multidisciplinary team can also involve paramedical staff of physiotherapists, speech and language therapists and occupational therapy. Again, the nature of the input from the different specialties is on a case-by-case basis and differs in every patient affected with cerebral palsy.

9.4.3 Corticobulbar Tract

The corticobulbar tract is a two-neuron path which unites the cerebral cortex with the cranial nerve nuclei in the brainstem involved in motor functions (apart from the oculomotor nerve). The following summary points will highlight the key facts in relation to the corticobulbar tract. This will allow quick and easy access to the key information relevant to these pathways.

(1) The corticobulbar tract originates in the ventral precentral gyrus
(2) The tract then passes through the internal capsule with the corticospinal tract
(3) Some fibers also pass in the tegmentum of the medulla and pons
(4) As the corticobulbar tract descends, it gives off branches to the motor nuclei of the trigeminal, facial, vagus, hypoglossal and spinal accessory nerves

TIP!

The motor nuclei of muscles involved in swallowing, speech, chewing and lingual movements are dual innervated, i.e. both cerebral hemispheres supply each of the sets of nuclei.

> **TIP!**
>
> The lower facial nuclei and the hypoglossal nerves are only innervated by one side (opposite cerebral cortex)

9.5 CLINICAL ASSESSMENT

Clinical assessment of the corticobulbar pathway will mean examining the cranial nerves which are supplied by this tract, i.e. trigeminal, facial, vagus, hypoglossal and spinal accessory nerves. This has been described in Chapter 1 where detailed accounts are given for each nerve. For further information related to clinical testing and pathologies, consult the companion text to this book – Clinical Anatomy of the Cranial Nerves (2014).

9.6 PATHOLOGIES

9.6.1 Pseudobulbar palsy

Pseudobulbar palsy is disease affecting the corticobulbar tracts bilaterally. The pathology can affect anywhere along the course of the corticobulbar tracts from the origin in the cortex, to its termination in the cranial nerve nuclei (trigeminal, facial, vagus, hypoglossal and spinal accessory nerves).

Pseudobulbar palsy can be caused by a variety of pathological conditions including trauma, neurological disease (Parkinson's, multiple sclerosis, etc), metabolic, vascular or tumor. Any condition which affects the corticobulbar tracts bilaterally will result in a pseudobulbar palsy.

It typically presents with difficulty with deglutition and mastication, increase of tongue reflexes and spasticity not only of the tongue but also the musculature of the pharynx and larynx. Speech will also be affected and has been thought of as resembling Donald Duck.

The patient will be unable to move their palate, and may have difficulty swallowing (dysphagia) and may also be emotionally inappropriate.

The investigation of these patient's must involve a neurologists who may perform neuroimaging (CT/MRI) as relevant and will investigate the cause of the palsy. Treatment and prognosis depends on what the underlying cause may be.

9.7 EXTRAPYRAMIDAL TRACTS

Extrapyramidal tracts are a network of nerves which is responsible for the motor control of activities like involuntary movements and reflexes. It is also concerned with coordination, i.e. motor control of movement. Rather, the extrapyramidal tracts work not by specifically targeting motor neurons, but rather work by modulating and regulating them.

The extrapyramidal tracts are so called to differentiate them from the pyramidal tracts (corticobulbar and corticospinal tracts) which pass through the pyramids of the medulla. The extrapyramidal tracts are found in the reticular formation in the medulla and pons. Their target neurons are found in the spinal cord and are responsible for movement, walking and reflexes. These tracts are influenced by other pathways from the basal ganglia, sensory cortex, vestibular nuclei and also the cerebellum. Therefore, they influence motor activity without directly innervating the motor neurons within the ventral gray horn.

The extrapyramidal tracts are the vestibulospinal, olivospinal, reticulospinal, rubrospinal and tectospinal tracts. These will now be discussed in turn.

9.7.1 Vestibulospinal Tract

The vestibulospinal tract is comprised of a lateral and medial pathway. The function of these tracts is to maintain equilibratory reflexes from the input of the vestibular apparatus. They will reach the axial muscles, i.e. intercostal and back muscles, as well as the extensors of the limbs.

9.7.1.1 Lateral Vestibulospinal Tract

The lateral vestibulospinal tract commences in the *lateral vestibular nucleus (Deiter's nucleus)* at the level of the pons and medulla. It descends through the medulla and passes into the ventrolateral and ventral regions of the spinal cord white matter. They terminate in the ventral gray horn. This tract is uncrossed.

9.7.1.2 Medial Vestibulospinal Tract

The medial vestibulospinal tract commences in the *medial vestibular nucleus (Schwalbe's nucleus)* in the pons and medulla. The fibers then pass into the medial longitudinal fasciculus ipsilaterally and contralaterally.

This pathway terminates on laminae VII and VIII neurons. This tract innervates the muscles which are involved in supporting the head and therefore only pass to the cervical region.

9.8 CLINICAL ASSESSMENT

The *vestibulospinal reflex* is where the vestibular apparatus interacts with skeletal muscle in maintaining posture and balance as well as ensuring the body is stable. This is demonstrated when the head moves resulting in stimulation of the *semicircular canals* as well as the *saccule* and *utricle*. This activates the vestibular nuclei via the vestibular nerve. Impulses are conducted through the vestibulospinal tracts to the spinal cord stimulating neck muscles. There are two types of reflex which are involved in maintaining normal body and head positions.

9.8.1 Righting Reflex

The purpose of this reflex is to return the head into the "normal" position. The input of this reflex is from the sensory, visual and vestibular systems and is classically highlighted in the cat when it falls and orientates itself on landing.

9.8.2 Tonic Labyrinthine Reflex

This reflex is formed from the position of the head and the effect of gravity. It has two components – supine and prone. The supine tonic labyrinthine reflex is when, during the first and second month after birth, the child is placed on their back and there is pulling of their head, trunk, upper and lower limbs toward the ground. The prone tonic labyrinthine reflex is for developing muscles used in flexion that will balance the extensor muscles. This must develop in order for the child to lift their body from the ground when lying on the stomach. Both of these reflexes will disappear by the fourth month, and if they do not, the child will have difficulty rolling from their back to the stomach and vice versa (Cheatum and Hammond, 2000).

Assessment of the vestibulocochlear nerve has been discussed in Chapter 1 and further details of clinical examination and pathologies of this nerve can be found in the companion text to this book (Rea, 2014)

9.8.3 Olivospinal Tract

Information in the olivospinal tract was initially thought to originate from the inferior olivary nucleus and crossover at that point, and then onward to the motor neurons within the ventral gray column, however, that has now been questioned and is doubtful if it actually exists (Snell, 2009). Remember here that the inferior olivary nucleus does, however, receive input primarily from the cerebellum, and may exist also with the cerebral cortex.

9.8.4 Reticulospinal Tract

The reticulospinal tract arises from the reticular formation and has three specific tracts all involved in different functions.

One group of descending fibers from the reticular formation controls autonomic functions and arises from the ventrolateral medulla. These fibers will project to the thoracolumbar spinal cord at the intermediolateral column. These neurons in turn will provide sympathetic innervation to the visceral organs.

Another of the reticulospinal tracts emerges from the medulla from the nucleus gigantocellularis. Neurons from this nucleus project bilaterally to all segments of the spinal cord. This pathway is referred to as the *lateral reticulospinal tract*, or *medullary reticulospinal tract*. This pathway reduces extensor spinal reflex pathways significantly. In the medial pontine reticular formation, another two groups of nuclei are found – the *nucleus reticularis pontis oralis* and the nucleus *reticularis pontis caudalis*. This is the medial reticulospinal tract and these ipsilaterally projecting neurons facilitate extensor responses.

Finally, there is a group of descending fibers which pass from the periaqueductal gray matter to the nucleus raphe magnus. The next part of this pathway links to the dorsal horn of the spinal cord, which then synapse onto primary afferent pain fibers. These fibers and pathways provide modulation of pain.

9.8.5 Rubrospinal Tract

In humans, the rubrospinal tract is very small. A small bundle of fibers from the red nucleus on the contralateral side continue as the rubrospinal tract. It is found ventral to the lateral corticospinal tract and terminates in the upper cervical segments of the spinal cord. This suggests that it has some function In relation to the upper limbs, and not the lower limbs. It has been suggested that it has a role in motor function. It is suggested that specifically, the main role is to aid motor neurons which innervate the flexor muscles (Siegel and Sapru, 2010). It should be noted that the red nucleus, part of the midbrain, has a major input from the primary and premotor cortex, therefore will control motor activity. Although in humans, the rubrospinal tract is rather small, the descending path from the red nucleus to the cerebellum is larger thus proving that the red nucleus has more of an influence on the cerebellum in primates (Siegel and Sapru, 2010).

9.8.6 Tectospinal Tract

The tectospinal tract originates from the superior colliculus, which itself receives information from the retina and cortical visual association areas. The axons in this pathway descend around the periaqueductal gray matter and then cross the midline at the point called the *dorsal tegmental decussation*. They then join with the medial longitudinal fasciculus within the medulla and pass in the ventral funiculus of the spinal cord. They terminate on the neurons within laminae VI–VII. The tectospinal tract is responsible for controlling the movement of the head in response to auditory and visual stimuli. Therefore, it has been postulated to control postural change on the visual information received to the superior colliculus.

9.9 CLINICAL ASSESSMENT

It is difficult to test the tectospinal tract in isolation due to the mixture of auditory and visual information it processes, but can include examination of those pathways from the respective cranial nerves influencing them, i.e. optic, oculomotor, trochlear, abducent and vestibulocochlear nerves, as well as the motor function of the muscles of the neck.

9.9.1 Dorsal Longitudinal Fasciculus

The dorsal longitudinal fasciculus is found within the dorsal brainstem tegmentum. It passes through the periaqueductal gray matter and contains both ascending and descending fibers.

The ascending fibers pass from the reticular formation passing to the hypothalamus thus transmitting information related to the viscera.

In turn, the descending portion arises also from the hypothalamus and passes to a variety of brain areas responsible for processing pain, cardiorespiratory functions and the autonomic system. Finally, the efferent fibers will also terminate on the preganglionic fibers of the autonomic nervous system.

9.9.2 Medial Longitudinal Fasciculus

The medial longitudinal fasciculus is found in the brainstem and is a set of crossed fibers with ascending and descending fibers.

The medial longitudinal fasciculus links the three main nerves which control eye movements, i.e. the oculomotor, trochlear and the abducent nerves, as well as the vestibulocochlear nerve. The purpose of the medial longitudinal fasciculus is to integrate movement of the eyes and head movements. It also forms a major component of the optokinetic and vestibule-ocular reflexes. The optokinetic reflex is when a patient follows an object and there are both smooth and saccadic eye movements. It allows following an object out of the field of vision and the eye returning to the original position it was when viewing the object initially. The vestibulo-ocular reflex is a reflex eye movement which results in stabilizing the image which falls onto the retina by moving the eye in the opposite direction to which the head moves.

Also within the medial longitudinal fasciculus are the vestibulospinal (medial) and tectospinal tracts, which innervate the neck musculature and the upper limbs.

The medial longitudinal fasciculus descending portion comes from the vestibular nucleus and is concerned with the gaze reflex. The input to the vestibular nucleus to allow this to occur is from the eight cranial nerve (vestibulocochlear nerve), fastigial nucleus and flocculus of the cerebellum. The fastigial nucleus processes information from proprioceptors in the head and neck and also the muscle spindles of the ankle. The flocculus is concerned with adjustment of walking, specifically concerned with the gait cycle.

Fibers also descend within the medial longitudinal fasciculus from the superior colliculus for integration of visual input. Also, within the descending fibers is information related to the accessory oculomotor nucleus for tracking of visual information as well as the pontine reticular formation for muscle tone of the extensors.

9.10 CLINICAL ASSESSMENT

As the medial longitudinal fasciculus carries information related to the oculomotor, trochlear and the abducent nerves, the following examination can be undertaken.

(1) *Extraocular eye muscle testing* – this is used to test the oculomotor, trochlear and abducens nerves (as all of these nerves supply the extraocular muscles and, therefore, are involved in movement of the eyeball)
   (a) Ask the patient to keep their head still during the examination
   (b) With their eyes only, they should follow the tip of your finger (or pen torch, or pencil, etc)
   (c) The examiner should then move the object in the horizontal plane from extreme left to extreme right.
   (d) This should be done SLOWLY
   (e) When at the extreme left or right, with the examining object, stop!
   (f) Observe for nystagmus
   (g) Then an H-shaped pattern with the examining object can be drawn in space, which the patient should follow WITH THEIR EYES ONLY
   (h) All movements should be done SLOWLY to assess for eye movements.
(2) Pupillary reflex
   This can be tested in two ways:
   (a) Ask the patient to follow your finger as you move it closer to their eyes. This will induce accommodation and therefore, pupillary constriction.
   (b) Swinging flashlight test. Using a pen torch, shine it into each pupil, observing for constriction of the pupil being tested, and also the opposite one.

   The swinging flashlight test can be used. This involves shining a pen torch into each pupil and observing both pupils for pupillary constriction. The purpose of this test is to identify if a relative afferent pupillary defect (RAPD) is present.

Assessment of the vestibulocochlear nerve is undertaken using basic hearing tests, Weber and Rinne's test and otoscopy. There are numerous advanced tests that can be used depending on clinical suspicion (Rea, 2014).

9.11 PATHOLOGIES

9.11.1 Internuclear Ophthalmoplegia

The conjugate gaze is controlled by the medial longitudinal fasciculus. Lesions of this pathway and the abducent nerve results in *internuclear ophthalmoplegia*. When you ask the patient to look to the unaffected eye, the affected eye shows only minimal adduction. On inspection of the contralateral eye (to the site of pathology), that eye will abduct but will have nystagmus. The patient will complain of diplopia (double vision) on looking toward the *unaffected* eye. In younger patients with multiple sclerosis, bilateral internuclear ophthalmoplegia can be found.

There is a rare syndrome referred to as Wall-Eyed Bilateral INternuclear Ophthalmoplegia (WEBINO syndrome). Here the patient will have a bilateral exotropia on primary gaze, bilateral internuclear ophthalmoplegia with impaired convergence (Chakravarthi et al., 2013). Many causes have been identified for this pattern, but the most common is infarction at the level of the midbrain (Chen and Lin, 2007).

If the lesion affects the abducent nerve nucleus or the paramedian pontine reticular formation, as well as the medial longitudinal fasciculus on the same side, it will result in conjugate horizontal gaze palsy in one direction and internuclear ophthalmoplegia in the other. This is referred to as the one and a half syndrome, typically caused by multiple sclerosis, brainstem stroke or tumor or arteriorvenous malformations at the level of the brainstem (Wall and Wray, 1983).

REFERENCES

Chakravarthi, S., Kesav, P., Khurana, D., 2013. Wall-eyed bilateral internuclear ophthalmoplegia with vertical gaze palsy. QJM. doi: 10.1093/qjmed/hct021 Found at: http://qjmed.oxfordjournals.org/content/early/2013/01/25/qjmed.hct021.full (accessed 09.01.2014).

Cheatum, B.A., Hammond, A.A., 2000. Physical activities for improving children's learning and behavior. A guide to sensory motor development. Human Kinetics Publishers, Champaign, IL, USA, ISBN 0880118741.

Chen, C.M., Lin, S.H., 2007. Wall-eyed bilateral internuclear ophthalmoplegia from lesions at different levels in the brainstem. J. Neuroophthalmol. 27, 9–15.

Compston, A., Coles, 2008. Multiple sclerosis. Lancet 372, 1502–1517.

Heesen, C., Mohr, D.C., Huitinga, I., Bergh, F.T., Gaab, J., Otte, C., Gold, S.M., 2007. Stress regulation in multiple sclerosis: current issues and concepts. Mult. Scler. 13, 143–148.

Kahan, S., Miller, R., Smith, E.G., 2009. In a page. Signs and Symtpoms, second ed. Lippincott Williams and Wilkins, Baltimore, USA.

Lublin, F.D., Reingold, S.C., 1996. Defining the clinical course of multiple sclerosis: results of an international survey. Neurology 46, 907–911.

Multiple Sclerosis Trust. http://www.mstrust.org.uk/information/aboutms/keyfacts.jsp (accessed 18.08.2014).

Rea, P., 2014. Clinical anatomy of the cranial nerves, first ed. Academic Press, Elsevier, San Diego, USA.

Rothwell, P.M., Charlton, D., 1998. High incidence and prevalence of multiple sclerosis in south east Scotland: evidence of a genetic predisposition. J. Neurol. Neurosurg. Psychiatry 64, 730–735.

Saatman, K.E., Duhaime, A.-C., Bullock, R., Maas, A.I.R., Valadka, A., Manley, G.T., 2008. Classification of traumatic brain injury for targeted therapies. J. Neurotrauma. 25 (7), 719–738.

Siegel, A., Sapru, H.N., 2010. Essential neuroscience. Lippincott Williams and Wilkins, Baltimore, Maryland. USA, ISBN 9780781783.

Snell, R.S., 2009. Clinical neuroanatomy, seventh ed. Lippincott Williams and Wilkins, ISBN-10: 0781794277.

Tataru, N., Vidal, C., Decavel, P., Berger, E., Rumbach, 2006. Limited impact of the summer heat wave in France (2003) on hospital admissions and relapse for multiple sclerosis. Neuroepidemiology 27, 28–32.

Tsang, B.K., Macdonell, R., 2011. Multiple sclerosis – diagnosis, management and prognosis. Aust. Fam. Phys. 40, 948–955.

Visser, E.M., Wilde, K., Wilson, J.F., Yong, K.K., Counsell, C.E., 2012. A new prevalence study of multiple sclerosis in Orkney, Shetland and Aberdeen city. J. Neurol. Neurosurg. Psychiatry 83, 719–724.

Wall, M., Wray, S., 1983. The one-and-a-half syndrome: a unilateral disorder of the pontine tegmentum – a study of 20 cases and review of the literature. Neurology 33, 971–978.

Yarnell, J, 2013. Epidemiology and disease prevention: a global approach, second ed. Oxford University Press, ISBN 978-0-19-966053-7.

CHAPTER 10

Brainstem Tracts

10.1 CORTICOPONTINE FIBERS

Corticopontine fibers are fibers which arise from all areas of the cerebral cortex, i.e. frontal, parietal, temporal and occipital lobes. However, the largest of these fibers arise from the frontal lobe. The purpose of the corticopontine fibers is a line of communication with the opposite cerebellum to allow for the coordination of planned motor functions, as it terminates in the deeper pontine nuclei.

The corticopontine fibers pass initially from the cerebral cortex (frontal lobe primarily) to terminate in the pontine nuclei. From here, they then project through the middle cerebellar peduncle to the opposite cerebellum via the pontocerebellar fibers. This pathway from cerebral cortex to pons to cerebellum is crucial in influencing the cerebellar function and integrity.

10.2 CLINICAL ASSESSMENT

Assessment of the function of the cerebellum is described as follows.

(1) Always introduce yourself to the patient (in any clinical examination or history taking) and state your position

(2) *Assess gait.* Ask the patient to walk from one side of the room (or examining area) to the other. If they normally use an aid to walking, they should be allowed to do so.

(3) *Heel to toe.* The patient should be asked to walk forward by placing one heel in front of the toes then switching to the opposite side and to keep walking in this fashion for a short distance

(4) *Romberg's test.* Further details are also found in Chapter 8.

    (a) Ask the patient to stand up with their feet together, arms by their side and eyes open.

    (b) Then, ask the patient to close their eyes for approximately 20–30 s.

Essential Clinical Anatomy of the Nervous System. http://dx.doi.org/10.1016/B978-0-12-802030-2.00010-8

(c) The patient may exhibit mild swaying which is normal.

(d) It is possible to repeat the test two times to help assessment. If the patient loses their balance, it is said that they have a positive Romberg's test, or Romberg's sign.

(5) Check for a *resting tremor* by having the patient place their arms and hands out straight.

(6) *Assess muscle tone and power* as discussed in detail in Chapter 8.

(7) *Check for dysdiadochokinesis.* Ask the patient to touch one dorsal surface of the hand with the palmar surface of the opposite hand. The opposite hand should then rotate to the dorsal surface of the opposite hand. This alternating palmar/dorsal surface onto the opposite hand should be repeatedly as rapidly as possible for the patient. Dysdiadochokinesis is the inability to undertake this rapid movement.

(8) *Finger to nose.* The patient should touch their nose then the examiners finger which is held in space. The examiner should move their examining finger and the patient should repeat the movement of touching their nose and the moving examiners finger.

(9) *Heel to shin test.* The patient should be asked to place the heel of one foot at the knee of the opposite leg. Then roll the heel down the front of the shin and back up. This should be repeated several times. Repeat this on the opposite side several times too.

10.3 NUCLEI OF CRANIAL NERVES

The trigeminal, facial, vestibulocochlear, glossopharyngeal, vagus and accessory nerves all originate at the level of the brainstem. These nerves carry a variety of types of fibers within them and each will be discussed below.

10.3.1 Trigeminal Nerve

The trigeminal nerve arises from the lateral aspect of the pons comprised of a large sensory root and a smaller motor root. The trigeminal nerve has three components – ophthalmic, maxillary and mandibular. It contains two types of fibers in it – those for muscles of mastication (branchial motor) and sensory to the face (general sensory). The branchial motor component supplies the temporalis, masseter and the lateral and medial pterygoid muscles. The sensory supply from the face comes

Table 10.1. Details of the Nuclei of the Trigeminal Nerve Including Its Input, Output and Related Functions of Those Nuclei

Trigeminal Nerve			
Nucleus	Input	Output	Function
Motor	Ipsilateral and contralateral primary motor cortices Sensory nucleus of trigeminal nerve	Muscles of mastication Tensor tympani Tensor veli palatini Mylohyoid Anterior belly of digastric	Motor information
Chief sensory	Primary afferent fibers	Ventral posteromedial nucleus of thalamus	Pain, temperature and light touch from head
Spinal tract and nucleus	Aδ and C fibers	Ventral posteromedial nucleus of thalamus	Pain and temperature
Mesencephalic	Muscle spindles Periodontal ligaments Temporomandibular joint	Reticular formation Cerebellum Motor nucleus	Non-conscious proprioception of the face (lower jaw) . Jaw jerk reflex

from its three major branches – the ophthalmic, maxillary and mandibular divisions. A summary of the functions of the trigeminal nerve has been previously described in Chapter 1.

The trigeminal nerve has four related nuclei which are described in Table 10.1.

10.4 CLINICAL ASSESSMENT

Two aspects of the trigeminal nerve should be assessed – the motor and sensory components. The motor component supplies the muscles of mastication and the sensory component supplies the majority of the face, apart from the angle of the mandible. The great auricular nerve, arising from the second and third cervical vertebrae, supplies the angle of the mandible.

(1) Always take a detailed history from the patient.
(2) ALWAYS tell the patient what you will be doing and what you expect them to do in helping elicit any signs and/or symptoms.
(3) Observe the skin over the area of temporalis and masseter first to identify if any atrophy or hypertrophy is obvious.
(4) Palpate the masseter muscles while you instruct the patient to bite down hard. Also note masseter wasting on observation. Repeat this with the temporalis muscle.

(5) Then, ask the patient to open their mouth with resistance applied by the examining clinician at the bottom of the patient's chin.

(6) To assess the stretch reflex (jaw jerk reflex), ask the patient to have their mouth half open and half closed. Place an index finger onto the tip of the mandible at the mental protuberance, and tap your finger briskly with a tendon hammer. Normally this reflex is absent or very light. However, for patients with an upper motor neuron lesion, the stretch reflex (jaw jerk reflex) will be more pronounced.

(7) Also ask the patient to move their jaw from side to side.

(8) Next, test gross sensation of the trigeminal nerve. Tell the patient to close their eyes and say "sharp" or "dull" when they feel an object touch their face. Allowing them to see the needle, brush or cotton wool ball before this examination may alleviate any fear. Using the needle, brush or cotton wool, randomly touch the patient's face with the object. Touch the patient above each temple, next to the nose and on each side of the chin, all bilaterally. You must test each of the territories of distribution of the ophthalmic, maxillary and mandibular nerves.

(9) Ask the patient to also compare the strength of the sensation of both sides. In other words ask them to state if they feel any differences between the left and right sides. If the patient has difficulty distinguishing pinprick and light touch, then proceed to check temperature and vibration sensation using the vibration fork. You can heat it up or cool it down in warm or cold water, respectively.

(10) Finally, test the corneal reflex (blink reflex). This is generally not done routinely, and should only be assessed if clinical suspicion indicates there may be a pathology involving the trigeminal nerve, as it can be a little uncomfortable for the patient. You can test it with a cotton wool ball rolled to a fine tip. Ask the patient to look at a distant object and then approaching laterally, touching the cornea (not the sclera) checking if the eyes blink. Repeat this on the opposite eye. If there is possible facial nerve pathology on the side that you are examining, it is imperative to observe the opposite side for the corneal reflex.

Some clinicians omit the corneal reflex unless there is sensory loss on the face elicited from the history or examination, or if cranial nerve palsies are present at the pontine level. It is best to ensure a complete clinical examination is undertaken, however, especially if there is a possible pathology of the trigeminal nerve.

10.4.1 Facial Nerve

The facial nerve has a short course within the cranial cavity after emerging from the junction between the pons and medulla just lateral to the root of the sixth nerve. The facial nerve (motor root and nervus intermedius), accompanied by the eighth cranial nerve (the vestibulocochlear nerve), enters the internal auditory meatus, traveling in a lateral direction through the petrous temporal bone. At the point it meets the cavity of the middle ear, it turns backward sharply forming a "knee-shaped bend". This is also where the sensory ganglion, the geniculate (genu, L. knee) ganglion, is found. On reaching the posterior wall of the middle ear it then passes inferiorly to exit the skull at the sylomastoid foramen. The facial nerve then enters the parotid gland, giving rise to its terminal branches for the facial muscles.

The facial nerve is composed of several different components that allow it to carry out its various functions. The divisions of the facial nerve functions are broken down into branchial (arising from the branchial/ pharyngeal arches) motor, parasympathetic and sensory (general *and* special) fibers. A summary of the functions of the facial nerve has been previously described in Chapter 1.

The nuclei related to the facial nerve are seen in Table 10.2.

Table 10.2. Details of the Nuclei of the Facial Nerve Including Its Input, Output and Related Functions of Those Nuclei

Facial Nerve			
Nucleus	Input	Output	Function
Facial motor	Corticonuclear fibers from both cerebral hemispheres for upper face Corticonuclear fibers from opposite cerebral hemispheres for lower face	Facial muscles	Muscles of facial expression
Nucleus of the solitary tract	Anterior two-thirds of the tongue	Ventral posteromedial nucleus of the thalamus	Taste from anterior two-thirds of the tongue
Superior salivatory nucleus	Hypothalamus	Submandibular salivary gland and lacrimal gland	Salivation and lacrimation

10.5 CLINICAL ASSESSMENT

The following presents a summary of the clinical testing of the facial nerve. It primarily focuses on the most important supply of the facial nerve from the clinical perspective i.e. its supply to the muscles of facial expression.

- The purpose of the facial nerve is to ensure functioning of the muscles of facial expression (Figure 1)
- Inspect the face at rest noting any asymmetry (e.g. drooping, sagging and even smoothing of the normal facial creases)
- Then:
  1. Ask the patient to raise their eyebrows, and
  2. Ask the patient to frown, and
  3. Ask the patient to show you their teeth

> **TIP!**
>
> **Do not** ask the patient to "smile" as they may be very worried about their signs and symptoms / clinical condition, and may feel uncomfortable being asked to smile! On the other hand, make sure they have their own teeth/substitutes in place to prevent embarrassment when undertaking this examination.

  4. Ask the patient to close their mouth and to puff out their cheeks against tightly closed lips
  5. Scrunch up their eyes, and the examining clinician should then try to open them on behalf of the patient

> **TIP!**
>
> **Do** tell them what you are about to do, as the patient will feel surprised that you are trying to prise their eyes open! This ensures the patient does not feel uncomfortable during the examination.

- The purpose of the examination is to note asymmetry of the face, but also, to assess the strength (or otherwise) of the power of the facial muscles

10.5.1 Vestibulocochlear Nerve

The eighth cranial nerve is the vestibulocochlear nerve. It arises from the brainstem between the pons and medulla, and has two nerves

Table 10.3. Details of the Nuclei of the Vestibulocochlear Nerve Including Its Input, Output and Related Functions of Those Nuclei

Vestibulocochlear Nerve			
Nucleus	Input	Output	Function
Vestibular related			
Superior vestibular	Vestibular nerve	Oculomotor nerve Trochlear nerve	
Medial vestibular	Vestibular nerve	Medial longitudinal fasciculus and cervical cord	Head and neck movements Eye movements
Inferior vestibular	Vestibular nerve	Ventral spinal cord Oculomotor nerve Trochlear nerve	Eye and head movements
Deiter's	Vestibular nerve	Ventral horn of spinal cord Motor nuclei of oculomotor, trochlear and abducens nerves	Equilibrium Movement of eyes in relation to head
Cochlear related			
Ventral cochlear	Auditory nerve fibers	Superior olive Inferior colliculus Lateral lemniscus	Processing of auditory information
Dorsal cochlear	Auditory nerve	Superior olive Inferior colliculus	Complex auditory information

responsible for different functions. The vestibular nerve deals with information related to equilibration, as it is distributed to the saccule and utricle, as well as to the ampullary crests of the semi-circular ducts. The cochlear nerve deals with hearing, and is distributed to the hair cells of the spiral organ. On exiting the brainstem, it then passes into the petrous temporal bone via the internal auditory meatus, very closely related to the facial nerve. A summary of the functions of the vestibulocochlear nerve has been previously described in Chapter 1 (which is also summarised in Table 10.3).

10.6 CLINICAL ASSESSMENT

Testing of the vestibulocochlear nerve can be complex and it may be indicated that the patient may need assessment by a neurologist or audiological specialist. However, testing can initially be undertaken at the bedside to provide a basic knowledge about the integrity of the vestibulocochlear nerve.

10.6.1 Testing At the Bedside
10.6.1.1 Basic Testing

Assess hearing by instructing the patient to close their eyes and to say "left" or "right" when a sound is heard on the examined side. Vigorously rub your fingers together very near to, yet not touching, each ear and wait for the patient to respond each time they hear something. After this test, ask the patient if the sound was the same in both ears, or louder or duller in either one or the other ear. If there is lateralization or hearing abnormalities perform the Rinne and Weber tests using the 512 Hz tuning fork.

10.6.1.2 Weber Test

The Weber test is a test for lateralization. Tap the tuning fork strongly on your palm and then press the butt of the instrument on the top of the patient's head in the midline and ask the patient where they hear the sound. Normally, the sound is heard in the center of the head or equally in both ears. If there is a conductive hearing loss present, the vibration will be louder on the side with the conductive hearing loss. If the patient does not hear the vibration at all, attempt again, but press the butt harder on the patient's head.

10.6.1.3 Rinne Test

The Rinne test compares air conduction to bone conduction. Tap the tuning fork firmly on your palm and place the butt on the mastoid eminence firmly. Tell the patient to say "now" when they can no longer hear the vibration. When the patient says "now", remove the butt from the mastoid process and place the U of the tuning fork near the ear without touching it.

Tell the patient to say "now" when they can no longer hear anything. Normally, one will have greater air conduction than bone conduction and therefore hear the vibration longer with the fork in the air. If the bone conduction is the same or greater than the air conduction, there is a conductive hearing impairment on that side. If there is a sensorineural hearing loss, then the vibration is heard substantially longer than usual in the air.

Make sure that you perform both the Weber and Rinne tests on both ears. It would also be prudent to perform an otoscopic examination of both eardrums to rule out a severe otitis media, perforation of the tympanic membrane or even occlusion of the external auditory meatus, which all may confuse the results of these tests. If hearing loss is noted, an audiogram is indicated to provide a baseline of hearing for future reference.

10.6.1.4 Otoscopy

This will allow the external auditory meatus and the middle ear to be assessed by examination of the tympanic membrane.

Advanced testing of the vestibulocochlear nerve can be undertaken by referral to an audiological specialist, perhaps in consultation with the neurological team or the ear, nose and throat surgeons. Specifically, advanced testing can be undertaken using automated otoacoustic emission (AOAE), automated auditory brainstem response (AABR), pure tone audiometry (PTA) or bone conduction tests. These tests will assess the hearing element of the vestibulocochlear nerve.

Regarding the vestibular component of the vestibulocochlear nerve, rotation testing, electronystagmography, computerized dynamic posturography (CDP) or vestibular evoked myogenic potential (VEMP) recording may be indicated.

10.6.2 Glossopharyngeal Nerve

The glossopharyngeal nerve is the eighth cranial nerve. It arises as three or four rootlets at the level of the medulla oblongata. It passes out from between the inferior cerebellar peduncle and the olive, superior to the rootlets of the vagus nerve. It then sits on the jugular tubercle of the occipital bone. It then runs to the jugular foramen, passing through the middle part of it. At the point of entry to the jugular foramen, two ganglia are found – an inferior and superior ganglion. Both of these ganglia contain the cell bodies of the afferent fibers contained within the glossopharyngeal nerve. On passing through the jugular foramen, the glossopharyngeal nerve then passes between the internal carotid artery and the internal jugular vein, descending in front of the artery. It then passes deep to the styloid process and related muscles attaching on to this bony prominence. It then winds round the stylopharyngeus, passing deep to the hyoglossus and going between the superior and middle pharyngeal constrictors. A summary of the functions of the glossopharyngeal nerve has been previously described in Chapter 1 (which is also summarised in Table 10.4).

Table 10.4. Details of the Nuclei Related to the Glossopharyngeal Nerve Including its Input, Output and Related Functions of Those Nuclei

Glossopharyngeal Nerve			
Nucleus	Input	Output	Function
Spinal tract and nucleus	Aδ and C fibers	Ventral posteromedial nucleus of thalamus	Pain and temperature
Nucleus ambiguus	Corticobulbar tract	Motor fibers of the vagus nerve	Innervation of the soft palate, pharynx and larynx
Inferior salivatory nucleus	Parasympathetic input	Parotid gland	Salivation
Nucleus of solitary tract	Afferents for gag reflex	Nucleus ambiguus	Gag reflex

10.7 CLINICAL ASSESSMENT

The following can be undertaken when examining the integrity of the glossopharyngeal nerve in the clinical setting, and can be undertaken at the bedside, or examining room.

(1) Examining the glossopharyngeal nerve is difficult. Assessing it on its own is not possible, and an isolated lesion of this nerve is almost unknown. When assessing the glossopharyngeal nerve, the fist thing to do is simply *listening to the patient talking*. Any abnormality of the voice, e.g. hoarse, whispering or a nasal voice may give a clue as to an abnormality. Also, ask the patient if they have any difficulty in swallowing. The result of a glossopharyngeal nerve (and related cranial nerves, e.g. vagus and accessory nerves due to their close proximity to each other) may be dysphagia (difficulty swallowing), aspiration pneumonia or dysarthria (difficulty in the motor control of speech).

(2) To assess the function of the glossopharyngeal nerve (and also the vagus nerve) ask the patient to say "ahhhh" (without protrusion of their tongue) for as long as they possibly can. Do not, however, prolong this examination beyond what is necessary. Normally, the palate should rise equally in the midline. The palate (uvula) will move *away* from the side of the lesion if there is a problem with the glossopharyngeal (and perhaps vagus) nerve, i.e. toward the "normal", or unaffected side.

(3) Damage to the glossopharyngeal (and also the vagus) nerve, e.g. because of a stroke, may result in the loss of the gag reflex.

ALWAYS tell the patient what you will do before assessing the gag reflex, as it is not a pleasant examination, and may not always be necessary.
(4) A swab can be used to gently touch the palatal arch on the left then right hand sides. Try to assess the normal side first if you suspect a pathology.

10.7.1 Vagus Nerve
The vagus nerve arises from the medulla. It passes toward the jugular foramen between the glossopharyngeal and spinal accessory nerves. The vagus nerve has two ganglia associated with it – the superior and inferior ganglia. The vagus nerve then passes inferiorly in the carotid sheath between the internal jugular vein and the internal and external carotid arteries. As it descends, it is related to the internal jugular vein and the common carotid artery. Then, the right and left vagus nerves take different anatomical pathways.

On the right side, the vagus nerve passes anterior to the right subclavian artery and posterior to the superior vena cava. At the point where it is closely related to the subclavian artery, it gives off its recurrent laryngeal branch. This branch passes under the artery then posterior to it. It then ascends between the trachea and esophagus, both of which it supplies at that point. The right recurrent laryngeal nerve then passes closely related to the inferior thyroid artery. It enters the larynx behind the cricothyroid joint and deep to the inferior constrictor. The recurrent laryngeal nerve conveys sensory information from below the level of the vocal folds, and all of the muscles of the larynx on that side, except cricothyroid.

The left vagus nerve descends toward the thorax passing between the common carotid and subclavian arteries, passing posterior to the brachiocephalic vein. It gives off branches here to the esophagus, lungs and heart. It then passes to the left side of the arch of the aorta. From here, the recurrent laryngeal nerve is given off which descends underneath the arch of the aorta to ascend in the groove between the esophagus and trachea. As it does so, it gives off branches to the aorta, heart, esophagus and trachea. A summary of the functions of the vagus nerve has been previously described in Chapter 1 (which is also summarised in Table 10.5).

Table 10.5. Details of the Nuclei Related to the Vagus Nerve Including the Input, Output and Related Functions of Those Nuclei

Vagus Nerve			
Nucleus	Input	Output	Function
Nucleus ambiguus	Corticobulbar tract	Motor fibers of the vagus nerve	Innervation of the soft palate, pharynx and larynx
Dorsal nucleus of vagus nerve	Nucleus of solitary tract Hypothalamus	Parasympathetic to viscera	Parasympathetic innervation of viscera, e.g. gastrointestinal tract, lungs
Nucleus of solitary tract	Epiglottis Aortic body Viscera	Hypothalamus Amygdala	Processing of visceral afferent information
Chief sensory nucleus of trigeminal nerve	Primary afferent fibers	Ventral posteromedial nucleus of thalamus	Pain, temperature and light touch from head and laryngeal mucosa

10.8 CLINICAL ASSESSMENT

Testing of the vagus nerve is done in exactly the same way that the glossopharyngeal nerve is examined.

(1) When assessing the vagus nerve, as with the glossopharyngeal nerve, the fist thing to do is simply *listening to the patient talking*. Any abnormality of the voice, e.g. hoarse, whispering or a nasal voice may give a clue as to an abnormality. Also, ask the patient if they have any difficulty in swallowing.

(2) To assess the function of the vagus nerve (and the glossopharyngeal nerve) ask the patient to say "ahhhh" (without protruding their tongue) for as long as they can. Normally, the palate should rise equally in the midline. The palate (uvula) will move *away* from the side of the lesion if there is a problem with the vagus (and glossopharyngeal) nerve, i.e. pulling of the palate to the normal side. This is due to the pull of the musculature on the unaffected side.

(3) The gag reflex can also be assessed if relevant. You MUST tell the patient what you will do before doing this test, as it is unpleasant.

(4) Using a swab, GENTLY touch each palatal arch in turn, waiting each time for the patient to gag. Again, do not prolong this examination as it can be unpleasant for the patient.

> **TIP!**
>
> Vagus nerve pathology could present with the following, affecting one or all of its branches:
>
> 1. Palatal paralysis (absent gag reflex)
> 2. Pharyngeal/laryngeal paralysis
> 3. Abnormalities with the autonomic innervation of the organs it supplies (i.e. heart, stomach (gastric acid secretion/emptying), gut motility
>
>   Glossopharyngeal nerve pathology on the other hand will affect the following:
>
> 1. Dysphagia
> 2. Impaired taste and sensation on the posterior one-third of the tongue
> 3. Absent gag reflex
> 4. Abnormal secretions of the parotid gland, though difficult to assess from the patient accurately

10.8.1 Accessory Nerve

The accessory nerve has two roots – a cranial and spinal division. The cranial root arises from the inferior end of the nucleus ambiguus and perhaps also from the dorsal nucleus of the vagus nucleus. The fibers of the nucleus ambiguus are connected bilaterally with the corticobulbar tract (motor neurons of the cranial nerves connecting the cerebral cortex with the brainstem). The cranial part leave the medulla oblongata as four or five rootlets uniting together, and then to join with the spinal part of the accessory nerve just as it enters the jugular foramen (Figure 52). At that point, it can send occasional fibers to the spinal part. It is only united with the spinal part of the accessory nerve for a brief time before uniting with the inferior ganglion of the vagus nerve. These cranial fibers will then pass to the recurrent laryngeal and pharyngeal branches of the vagus nerve, ultimately destined for the muscles of the soft palate (not tensor veli palatini (supplied by the medial pterygoid nerve of the mandibular nerve).

The spinal root arises from the spinal nucleus found in the ventral gray column extending down to the fifth cervical vertebral level. These fibers then emerge from the spinal cord arising from between the ventral and dorsal roots. It then ascends between the dorsal roots of the spinal nerves entering the cranial cavity through the foramen magnum posterior to the vertebral arteries. It then passes to the jugular foramen, where it may receive some fibers from the cranial root. As it then exits the jugular foramen, it is closely related to the internal jugular vein. It then courses inferiorly passing medial to the styloid process and attached stylohyoid. It also is found medial to the posterior belly of digastric. The spinal root then supplies the sternocleidomastoid muscle on its medial aspect.

The cranial root then enters the posterior triangle on the neck lying on the surface of the levator scapulae at approximately mid-way down the sternocleidomastoid. As it passes inferiorly through the posterior triangle of the neck, and just above the clavicle, it then enters the trapezius muscle on its deep surface at its anterior border. The third and fourth cervical vertebral spinal nerves also supply the trapezius forming a plexus of nerves on its deeper surface.

The spinal accessory nucleus is formed from the lower motor neurons within the superior portion of the spinal cord in its dorsolateral aspect of the ventral gray horn. A summary of the functions of the accessory nerve has been previously described in Chapter 1.

10.9 CLINICAL ASSESSMENT

From the clinical perspective, the accessory nerve supplies the *sternocleidomastoid* and *trapezius* muscles, and as such, it is those that are tested when assessing the integrity of the nerve.

The sternocleidomastoid muscle has two functions depending on whether it is acting on its own, or with the opposite side. If the sternocleidomastoid is acting on its own, it tilts the head to that side it contracts and, due to its attachments and orientation, rotates the head so that the face moves in the direction of the opposite side. Therefore, if the left sternocleidomastoid muscle contracts, the face turns to the right hand side, and vice versa.

If, however, both sternocleidomastoid muscles contract, the neck flexes and the sternum is raised, as in forced inspiration.

The trapezius is an extremely large superficial muscle of the back. It is comprised of three united parts – superior, middle and inferior. It is involved in two main functions depending on if the scapula or the spine is stable. If the spinal part is stable, it helps move the scapula, and if the scapula is stable, it helps move the spine. Trapezius is involved in a variety of movements. The upper fibers raise the scapula, the middle fibers pull the scapula medially and the lower fibers move the medial side of the scapula down. Therefore, trapezius is involved in both elevation and depression of the scapula, depending on which part is contracting. As well as this, the trapezius also rotates and retracts the scapula.

Testing of the accessory nerve is done as follows.

(1) ALWAYS inform the patient of what you will be doing, after introducing yourself and taking a detailed clinical history.
(2) When examining a patient, ensure you just observe the patient in the resting position, and try to identify if there is any obvious deformity, or asymmetry of the shoulder and neck region. It may be that you will see an obvious weakness or asymmetrical position of the patient's neck and/or upper limbs
(3) First, you can assess the sternocleidomastoid muscle.
(4) You can ask the patient to rotate their head to look to the left and right hand sides to identify any obvious abnormality.
(5) Then, ask the patient to look to one side and test the muscle against resistance.
(6) For example, if the patient looks to the right side, place the ball of your hand on their left mandible.
(7) Ask the patient to press into your hand.
(8) Repeat this on the opposite side.
(9) Then, you need to assess the trapezius.
(10) First you can ask the patient to raise their shoulders, as in shrugging.
(11) Observe any gross abnormality
(12) Then while the patient is raising their shoulders, gently press down on them as they lift their shoulders.
(13) Assess any weakness that may be present, recording which side is affected.

> **TIP!**
>
> When assessing the function of the sternocleidomastoid and trapezius, it may help examining the unaffected side first, especially if the patient complains of pain or discomfort on one side. This helps build up trust with the patient, but also minimizes causing them any pain or discomfort.

10.10 INFERIOR OLIVARY COMPLEX

The inferior olivary complex, or inferior olive nucleus is found at the level of the rostral medulla. It is a swelling at the lateral aspect of the pyramids. As such, it is also closely related to the cerebellum. It is composed of gray folded layers opening medially by a hilum. At the point of the hilum, the olivocerebellar fibers pass through.

The purpose of the inferior olivary nucleus, due to its close relation to the cerebellum, is related to movement and the control and coordination of it. The inferior olivary nucleus is the only source of climbing fibers to the Purkinje cells in the cerebellum and it projects to both the cortex and the deeper nuclei of the cerebellum. It is also involved in processing of sensory information and tasks of cognition. It has also been implicated in the vestibule-ocular reflex and eye blinking (De Zeeuw et al., 1998).

10.11 OLIVOCEREBELLAR FIBERS

The cerebellum is involved in coordination and regulation of motor control, control of balance and posture, processing of vestibular information, cognitive functions (Goldman-Rakic, 1996; Schmahmann and Caplan, 2006) and has connections with the hypothalamus for autonomic and emotional functions (Schmahmann and Caplan, 2006).

The inferior olive provides all of the climbing fibers to the Purkinje cells, which are the only source of the output of the cerebellar cortex reaching central cerebellar and vestibular nuclei. The inferior olive receives sensory input following motor commands being executed, but also receives information from deep nuclei of the cerebellum and also the mesodiencephalic junction.

This pathway aids in learning and timing. The timing hypothesis as a function of the olivocerebellar function has been discussed by De Zeeuw et al. (1998) showing that the inferior olive functions as an oscillating clock, allowing for the correct timing of motor functions due to the nature of its neuronal electrophysiological characteristics.

REFERENCES

De Zeeuw, C.I., Simpson, J.I., Hoogenraad, C.C., Galjart, N., Koekkoek, S.K.E., Ruigrok, T.J.H., 1998. Microcircuitry and function of the inferior olive. Trends Neurosci. 21, 391–400.

Goldman-Rakic, P., 1996. Regional and cellular fractionation of working memory. Proc. Natl. Acad. Sci. USA 93, 13473–13480.

Schmahmann, J.D., Caplan, D., 2006. Cognition, emotion and the cerebellum. Brain 129 (2), 290–292.

INDEX

Printed in the United States
by Baker & Taylor Publisher Services